Aspects of African Biodiversity

Proceedings of the Pan Africa Chemistry Network Biodiversity Conference, Nairobi, 10-12 September 2008

Aspects of African Biodiversity
Proceedings of the Pan Africa Chemistry Network Biodiversity Conference, Nairobi, 10-12 September 2008

Edited by

J. O. Midiwo
Department of Chemistry, University of Nairobi, Nairobi, Kenya

J. M. Clough
Syngenta, Bracknell, UK

RSC Publishing

Special Publication No. 321

ISBN: 978-1-84755-948-7

A catalogue record for this book is available from the British Library

Published by The Royal Society of Chemistry,
Thomas Graham House, Science Park, Milton Road,
Cambridge CB4 0WF, UK

Registered Charity Number 207890

For further information see our web site at www.rsc.org

Preface

This book is a collection of papers based on the lectures presented at the PACN Biodiversity Conference held at the University of Nairobi, Kenya, on 10-12 September 2008.

The PACN, launched in November 2007 in London, UK, established hubs in Nairobi in March 2008 and in Addis Ababa, Ethiopia, in September of the same year. Both hubs are jointly funded by the Royal Society of Chemistry (RSC) and by Syngenta, a world-leading agribusiness. The PACN also extends to include RSC events and projects in South Africa and other regions of the continent.

The PACN aims to improve collaboration between governments, universities, industry and communities; to champion the successes of African science, attracting students to scientific careers and promoting public appreciation of the important role for chemical science in Africa's future; to support science education at all levels, from school teaching to training researchers in the use of modern instruments; and to provide information to researchers about funding opportunities.

One of the key activities of the PACN is to connect African scientists with each other and the rest of the world through a cycle of conferences and workshops. The Biodiversity Conference, held at the University of Nairobi from 10-12 September 2008 and organised by the Kenyan hub, was the first of these. It brought together 150 delegates from 12 African countries – Botswana, Burundi, the Democratic Republic of Congo, Ethiopia, Kenya, Nigeria, Rwanda, South Africa, Sudan, Tanzania, Uganda and Zimbabwe – as well as from Brazil and the UK.

This was the first international conference on the chemical sciences ever to be held on the beautiful campus of the University of Nairobi. The Conference Centre provided excellent facilities, and these were supplemented by tents set up on the lawns, from which refreshments and meals were provided. Thirty chemistry undergraduates from the University of Nairobi also attended and brought a fresh and lively presence to the conference. Nine of them, easily identified by their RSC T-shirts, helped with a variety of tasks throughout the conference.

The lectures and extensive discussions covered topics such as the chemistry and biological activity of natural products; making the best use of the diversity of Africa's plants in agriculture and medicine whilst benefiting the communities from which they come; agricultural practices for Africa; genetic diversity in farm animals and crop plants; analysis and the environment; the protection of water supplies from pollutants; understanding pesticides in the African context; conservation; biofuels; and the processing and use of waste materials. Taken together, these lectures provided a valuable overview of many aspects of biodiversity in Africa, and they are now collected as the chapters in this book.

J.M. Clough and J.O. Midiwo

Contents

CHALLENGES AND OPPORTUNITIES FOR AFRICA: AN OVERVIEW OF KEY THEMES FROM THE PACN BIODIVERSITY CONFERENCE

S.C. Smith

Syngenta, Crop Protection Research Chemistry, Jealott's Hill International Research Centre, Bracknell, Berkshire RG42 6EY, UK

1 INTRODUCTION

A wall of three-dimensional tiles representing different micro-organisms provided an impressive backdrop in the Conference Centre of Nairobi University, an appropriate setting indeed for the inaugural Conference on Biodiversity of the Pan Africa Chemistry Network. This prologue chapter will give a flavour of the wide range of topics covered at the conference and will highlight some of the key themes that emerged from the individual scientific contributions. Much of the material was provided to the participants at the end of the conference in a 'rapporteur summary' and I would like to gratefully acknowledge the valuable contributions of my co-rapporteurs in distilling out the main points.

Firew Mekbib from Ethiopia defined biodiversity during one of the sessions as 'the totality of genes, species and ecosystems in a region'. The range of presentations at this conference reflected the diversity of science that impacts our understanding and practical utilisation of the rich biodiversity in Africa. They included areas such as: African and global plant biodiversity; the use of natural products in medicines and crop protection; biofuels; management and conservation of biodiverse ecosystems; genetic variation in crops and animals; environmental monitoring of pollutants; and waste management.

Syngenta has summarised its goal as *'bringing plant potential to life'*. In doing so, sustainable agriculture is of fundamental importance and this relies on maintaining the biodiversity of agricultural environments. As an example, Syngenta is a member of the Global Crop Diversity Trust, a foundation committed to conserving crop diversity for global food security. It was therefore of particular interest to see how many of the presentations touched on this theme, reflecting the importance of agriculture in Africa and the real challenges it faces in the future.

The conference began with a thought-provoking puppet show from the Chanuka Express Theatre Company, which highlighted some of the concerns of young Africans about the adverse impact man can have on our environment. The play focused on the indiscriminate use of pesticides, the problem of polluted water, and human actions upsetting the balance of nature. In reality, there is an enormous global challenge to increase agricultural productivity in a sustainable way, with limited land, to meet the requirements for food, feed, fibre and fuel of a growing and increasingly affluent population. The biodiverse ecosystems we currently have need to be maintained, while increasing the quality and quantity of production from existing land. A loss of biodiversity makes an ecosystem more vulnerable to changes in the environment. There may be fewer species to support fundamental ecosystem functions, such as pollination. Realistically, the agricultural challenge can only be achieved through technological innovation, which will include chemical pest control and biotechnology. Modern crop protection chemicals comply with strict environmental regulations and bring great benefits to agriculture. Unfortunately, the best products are not always readily available to African farmers who, instead, may have to rely on older or generic chemicals with the associated issues of regulation and stewardship.

Although the conference highlighted many threats to the environment and potential detrimental effects on biodiversity, we should recognise that these problems present important opportunities for science, and chemistry in particular, both to thoroughly understand the issues and then provide innovative solutions. There are some huge global challenges (such as food, water, energy and health), which were presented with an African dimension. There are also opportunities for innovation in technology and for improving our basic understanding of the underlying sciences. These should provide academics with important, and highly relevant, targets for their research and our young scientists with inspiration to develop their careers in the future. We should also start to feel that, as scientists, we can be driving potential solutions and counter the all-too-common public perception that science is part of the problem.

2 KEY THEMES FROM THE CONFERENCE

Sebsebe Demissew (National Herbarium, Addis Ababa University, Ethiopia) started the conference off in style by introducing the tremendous range of plant diversity in Africa, where there are 40-45,000 species growing across many diverse habitats, equating to over a fifth of the world's flora. We rely on plants for food, medicines, crop protection, and materials for clothing and building. They also provide ecological services, cleaning our air, water and soil. Wild relatives of cultivated species provide a pool of genetic diversity for future crop improvements. Plants also offer technical and commercial opportunities for Africa. Examples include low-caffeine coffee and the nutritious grain Tef from Ethiopia which are marketed in the West. The latter is a gluten-free crop that is used in health foods and may be of benefit to diabetics. There are some important challenges, which include unsustainable harvesting, burning, human interference, climate change and ensuring local communities benefit from the exploitation of both their natural resources and know-how. Last, but not least, there is a lack of new botanists being trained to help us understand the true untapped potential of African plants.

Africa contains two out of the five global 'higher biodiversity areas' and nine biodiversity hotspots. Mkabwa Manoko (University of Dar es Salaam, Tanzania) stressed the need to ensure conservation and restoration of Africa's biodiverse areas and for local communities to share in the benefits derived from any exploitation of biodiversity, particularly where this originated from indigenous knowledge of traditional medicines. Of the drugs prescribed in the USA, 56% have a natural origin, reflecting the importance of this resource. Threats to biodiversity include population growth, invasion by exotic species, destruction of natural habitats and changes in land use. Fundamental inequality between production and consumption is an issue facing Africa, and future shared benefits will rely on equitable partnerships with local communities.

Berhanu Abegaz described NABSA (Network for Analytical and Bioassay Services in Africa, University of Botswana) as an analytical centre of excellence and a model of networking in the discovery of new natural products. He outlined the approach used at NABSA for the identification of secondary metabolites from the huge plant diversity in Africa (sometimes *via* ethnobotanical approaches), as well as research into their properties, biosynthesis and function. NMR instruments operating at frequencies up to 600 MHz are used in this research, and are also available on a service basis for other African scientists. Importantly, this centre of excellence goes some way towards making up for labs elsewhere that are often poorly equipped, enabling African scientists to engage in short-term collaborations, publish in high-impact journals, and to begin to compete at an international level. It also encourages African scientists to work on African problems.

Jacob Midiwo (NAPRECA, University of Nairobi) then went on to discuss the opportunities that plant-derived remedies present for the treatment of neglected diseases in Africa. The WHO estimates that 80% of the world's population already rely on traditional plant-derived medicines. There are still opportunities for new therapies; for example, over 500 plants are potential sources of anthelmintics. The high levels of active components (up to 10-15% in some cases) offer the possibility of field treatments based on plant extracts. This would avoid the need for very expensive drug development, which is out of the reach of most African organisations. Malaria is a critical global problem with 300-500 million cases a year, 80% of which are in Africa. In Kenya alone, 22 million people are at risk, and children are particularly vulnerable. However, trying to control malaria is like 'shooting at a moving target' as resistance develops rapidly. African 'ethno-medicines' or 'Herbal Combination Therapies' (HCTs) based on plant extracts may offer an alternative to ACTs (Artemisinin Combination Therapies). Professor Midiwo described a bioassay-guided fractionation approach to the identification of novel compounds such as diprenylated flavanones. The next challenge is to develop novel formulation methods for these compounds. Lawrence Arot Manguro (Maseno University, Kenya) also used bioassay-guided fractionation in the isolation of sesquiterpenes from the African evergreen tree *Warburgia ugandensis* as potential antifungals.

Charles Wambebe (FAS, FTW, International Medical Research in Africa, Abuja, Nigeria) continued the theme of African medicinal plants. Of the new medicines introduced between 1981 and 2002, 67% are of plant origin. However, less than 10% of the ~70k African plants have been explored for their potential. Intellectual property rights must be considered and benefits shared equitably with local communities and scientists. This creates a need for better professional governance and ethics. Pharma pipelines can be long and expensive and are therefore unrealistic for Africa. There are issues for the development of herbal remedies which will contain mixtures – including safety, quality control (fingerprint profiling) and ethical review of clinical research (most countries do not have a review body at a national level). Despite these issues, there are some success stories, e.g. Niprisan from sorghum, which is licensed to the U.S. company XeChem as a sickle-cell treatment, with a share of the profits fed back to the African research institute.

Tony Hooper (Rothamsted Research, UK) discussed local applications of 'push-pull' insect-control strategies with plant-derived semiochemicals and pheromones in the control of crop pests. For example a sudan grass trap crop reduced the number of stem borers on maize grown in adjacent plots. Five thousand African farmers have already been involved in a practical test of the methodology which exploits plant diversity in a very creative way. It was an excellent example of an understanding of chemistry and biotechnology being put to practical use in a local context.

John Clough (Syngenta, UK) described the Syngenta approach to agrochemical discovery and, more specifically, the screening of natural products as potential leads and the synthesis of analogues. Syngenta has collaborated with external partners to acquire novel natural products, ensuring an equitable agreement to share in any successes. Rediscovery of known compounds is common and many natural products are only weakly active, unstable or have undesirable functionality for use as agrochemicals, so relatively few make it through into research programmes. Synthetic strategies take into account the physical properties of targets and scope for simplification. There are significant challenges in the optimisation of natural product leads to commercial agrochemicals, and in some cases this has required the synthesis of thousands of analogues, something that is only feasible within a commercial organisation. Syngenta makes more use of micro-organisms than plants as a source of natural products, since it is easier to re-supply larger quantities of compounds produced in this way. Africa must have a huge and diverse array of micro-organisms, and these surely represent a huge research opportunity for the future. Insects and arthropods make up over half of animal species and as such are relatively under-explored as sources of natural products. John Bwire Ochola [International Centre of Insect Physiology and Ecology (ICIPE), Kenya] has been using bioassay-guided fractionation of the venom from Kenyan scorpions in a search for selective insecticides.

Hayder Abdelgader (Agricultural Research Corporation, Sudan) talked about 'soft pesticides' – those commercial insecticides with a low impact on beneficial or predatory insects – and their role in integrated pest management in conjunction with natural enemies. This strategy recognises the vital importance of considering the wider agricultural ecosystem when controlling crop pests.

Of the ecosystems on which the world depends, 60% are being degraded. How do we access local knowledge to assess the potential impact on crop diversity and genetic erosion? Firew Mekbib (Harmaya University, Ethiopia) described the key roles farmers can play in the conservation of crop biodiversity in sorghum – known as the 'camel crop of cereals' because of its drought tolerance. Interestingly, sorghum farmers have developed their own 'folk taxonomy tree' and a local seed bank. This helps them to breed specific varieties for food, feed and materials and to very effectively contribute to long term food security *via* increasing genetic diversity. The discussion after Firew Mekbib's presentation highlighted Striga as a neglected African weed in sorghum.

Ethiopian dairy cattle genetics were the subject of the presentation by Henry Mutembei [International Livestock Research Institute (ILRI), University of Nairobi]. There is a rising demand for milk and meat (expected to more than double by 2020) so an understanding and optimisation of cattle genetics will be of great value. There are knock-on beneficial effects in tackling farmer poverty and relieving pressure on natural resources. An *in vitro* embryo production (IVEP) tool can be applied in cattle breeding to identify, select and conserve key traits. The ambition of ILRI's research is to tailor cattle genetics to specific environmental regions and thereby maximise outputs of milk. Tree biodiversity can also be monitored through genetics. J.S. Nantongo (Makerere University, Uganda) described genetic tools that would allow us to understand the gene flow and diversity of forests and thereby identify the most effective strategies for restoration and the conservation of biodiversity.

Biofuels were a major theme at the conference with local African plants offering potential sources of renewable fuel. Jeremy Woods (Imperial College, London, UK) gave an overview of some of the key issues. The high price of petroleum has been the key factor driving the development of biofuel production, but energy security, climate change and rural development are also important. Bioenergy offers the potential to supply 30-50% of projected primary energy globally by 2050. Dr Woods mentioned that to supply Africa with fuel costs $70bn a year. It is therefore a significant cost to many African economies and, for example, 40% of the foreign exchange earnings in Tanzania and Senegal were spent on oil products in 2007. True sustainability is the ultimate goal, but it is not trivial to quantify in order to compare different fuels as there are a number of contributing factors to consider, including water use; soil carbon and quality; waste materials and recycling; carbon stocks and biodiversity. Destruction of forest to grow biofuel crops has been shown to have the biggest negative environmental impact on sustainability scores. Countering this, utilisation of waste by-products can be a really positive factor and non-tillage-based agriculture is better than classical tillage. We also need to consider 'radiation-use efficiency' from one region to another. There is also a need to overcome a yield and productivity gap. Maize grown in the USA, for example, is much more productive than that grown in Zambia or Tanzania. On the positive side, land is not such a constraint in Africa as it is in some developed countries.

Paolo Cézar Vieira (Federal University of São Carlos, Brazil) also talked about biofuels, but focused specifically on biomass conversion. There are around 20,000 edible plants in the world but fewer than 20 provide 90% of the world's food. There are many other potential uses of plants as sources of a variety of raw materials including drugs, pigments, oils and fibre. Dr Vieira discussed the pros and cons of various renewable fuels – ethanol, butanol, biodiesel and hydrogen. Key issues will include the quality control of biofuel and the use of by-products to improve sustainability. There are many opportunities for new technologies, for example in the conversion of glycerol (a biodiesel by-product) into hydrogen. Future economic use of ethanol will depend on the efficient processing of lignocellulose. In the USA, 36 billion gallons of ethanol will be needed by 2020 and half of this will have to come from cellulosic sources. A key message was that the world needs to consume less in the future and should not expect new technologies to allow our current way of living to continue.

In a more specific African context, C.S. Shonhiwa (Finealt Engineering, Zimbabwe) highlighted the potential of jatropha as a source of biodiesel in Zimbabwe. Jatropha will grow on relatively poor quality land and still provides a high yield of oil. Zimbabwe has adequate land and human resources, and so this crop provides an attractive economic source of renewable fuel. To this end, two jatropha diesel plants are planned for 2010. Significant issues to address include setting up suitable infrastructure and storage; pest control in jatropha; monitoring and maintaining oil quality and the assessment of environmental impact.

Environmental monitoring is essential to determine the impact of pollution on biodiversity. Water is of fundamental importance to all ecosystems and this scarce resource is particularly susceptible to pollution. Coincidentally, Nairobi, the location of the conference means 'cool water' in the Masai language. Factors linked to water quality, or scarcity, include disease (cholera, diarrhoea and malaria, particularly in children), irrigation, deforestation, population (use and waste), climate change and pollution. Professor Shem Wandiga (University of Nairobi) focused on the monitoring of pollutants such as phosphorus, nitrogen and PCBs in Lake Victoria. These can be traced to a number of sources, but untreated sewage waste from the increasing population living around the lake appears to be the main one. The impact is clearly evident in the lake, where pollution, algal blooms, sedimentation and eutrophication (leading to choking by water hyacinth) and a reduction in water volume, are becoming more common, leading to a collapse in fish stocks. A detailed understanding *via* thorough chemical analysis should ultimately enable solutions to be put in place.

Shadrack Mule (University of Eastern Africa, Baraton, Kenya) has been measuring and monitoring contamination of water and sediments by heavy metals such as cadmium and lead stemming from the activities of small engineering and other businesses known as 'Jua Kali'. There is obviously a need to balance the economic growth and employment that

Jua Kali can bring in under-developed regions with possible long term environmental damage. E. Muleya (Midlands State University, Zimbabwe) has analysed the effluent from battery manufacturers, specifically the levels of zinc and cadmium. High levels of some metals were found in boreholes but, thankfully, not in tap water. Simple effluent treatment technology to remove heavy metals (for example *via* precipitation with lime) was proven to be effective and should improve the situation.

Michael Kishimba (ANCAP, SETAC, University of Dar es Salaam, Tanzania) reported results on the analysis of pesticide residues in food crops and the environment in Tanzania where 70% of the 38 million population are employed in agriculture. This type of analysis is complicated because of the range of compounds potentially in use. African farmers have a low purchasing power, leading to the continued widespread use of old agrochemicals. Disposal and dumping can become serious issues. A shocking example of this was presented, where 1400 tonnes of obsolete pesticides had contaminated a particular region of Tanzania. In addition, farmers may not be adequately trained, leading to improper handling and use of agrochemicals. Despite this, pesticide residue analysis in food generally gave acceptable results, although there was concern that limits may be not be set at the same strict levels as in the European Union. There was a vigorous discussion about the relative benefits of DDT in malaria control versus its negative environment impact.

Engdawork Admasu (Bahir Dar University, Ethiopia) has analysed nutrients and non-essential metals in cows' milk. Encouragingly, toxic heavy metals were generally found to be below acceptable levels. By contrast, lead levels were relatively high and could be attributed to the continued use of leaded petrol. Pollution by ammonia affects plant growth rates, crop yields and aquatic life. Crispin Kowenje (Maseno University, Kenya) has developed a novel zeolite methodology for the analysis of ammonia. There is great scope to develop other innovative new tools for environmental monitoring.

Steve Lancaster of FASTA (Foundation for Analytical Science and Technology in Africa, UK) provided an excellent example of the use of analytical science in the measurement, monitoring and control of a significant environmental problem, namely the negative effect of the anti-inflammatory drug diclofenac in the food chain of Gyps vultures in India. Through FASTA, Steve has been involved in the introduction and commissioning of GC-MS technology in Kenyan university chemistry departments. One of these, at Jomo Kenyatta University, has been used in the vulture case study to analyse samples from animal and bird carcasses to evaluate the extent of the problem and then to monitor subsequent compliance with restrictions on diclofenac use. This project facilitated introduction of new equipment; development of a new analytical capability and research methodology; provision of a monitoring service and the training of chemists, and so provides an excellent model for future initiatives in African universities.

Naftali Muriithi (Kenyatta University, Kenya) has recognised the potential value in municipal waste as a resource. With increasing population, waste is also increasing rapidly. However this waste is difficult to recycle because it tends to be very heterogeneous in nature. It may be a mix of domestic and industrial materials and can also contain medical waste. A specific opportunity is to make use of nutrients present in biomass ash from domestic waste. Key factors to make this work will include more proactive government policy and constructing the right infrastructure.

Zhou Leocadiah (University of Fort Hare, South Africa) emphasised the need for a coordinated approach to managing animals, water and tourism in complex biodiverse ecosystems, with an example from the Kruger National Park. The involvement of many stakeholders may be essential for conservation to be successful. The operation of the A.P. Leventis Ornithological Research Institute (APLORI) in Nigeria, described by Georgina

Mwansat, provided an excellent example of the provision of socio-economic and community benefits by effectively involving them in the monitoring of local biodiversity and habitats. Research centres of this type provide education of the next generation of scientists; development and training of local people, as well as a focus for setting up long-term ecological projects. Georgina Mwansat and her colleagues have recorded 883 bird species, and have used this high degree of bird diversity as an indicator of general biodiversity.

3 CONCLUSIONS AND OUTLOOK

One of the main goals of the Pan Africa Chemistry Network is to connect African scientists more effectively, both with each other and with the international chemistry community. The PACN Biodiversity Conference was a great first step towards this goal with 12 African countries represented, along with a significant number of participants from beyond the continent. All of those present will have gone away more aware of several critical challenges facing Africa, linked by a common biodiversity thread. These include sustainable agriculture, health and neglected diseases, long-term fuel supply and the environmental impact of human activity. These are serious issues, but they should also provide inspiration and a sense of urgency to African scientists as they direct their research towards solving these problems.

Members of the Nairobi University Chemical Club provided excellent support with logistics at the conference. It was great to see that they also attended the scientific sessions. These young people will have heard accounts of many significant global challenges, put into an African biodiversity context. I hope they have left the conference inspired to develop their careers in chemistry and to understand and address these challenges for the future benefit of their continent.

EXPLORING THE THERAPEUTIC POTENTIALS OF AFRICAN MEDICINAL PLANTS

C. Wambebe

International Biomedical Research in Africa, Number 4 Aso Drive, Abuja, Nigeria.

1 INTRODUCTION

Prior to the advent of colonialists to Africa, Africans depended entirely on African traditional medicines (ATM) for the management of all their health problems. The advent of the colonialists to Africa brought orthodox medicines to Africa resulting in a change of attitude of the elites to ATM. Generally, such Africans prefer orthodox medicines to ATM. The rapid effect of antimicrobial agents and modern diagnostic facilities as well as religious beliefs may be responsible for this change in attitude. Undoubtedly, modern science and technology has revolutionalized human health. Unfortunately, the impact of modern medicine however has not dramatically changed the health status of Africans. Consequently, about 70% of people living in the developing countries routinely use traditional medicine to treat their prevailing diseases.

According to the WHO, the criteria for evaluating traditional medicines, like those for conventional medicines, are quality (i.e. medicines should be well formulated and possess consistent quality from batch to batch as specified in the official pharmacopoeia chosen as a standard), safety (i.e. medicines should not present risks that are disproportionate to their benefits) and efficacy (i.e. medicines should be effective for the indications claimed and should be clinically significant and useful). Incidentally, most traditional medicines used in Africa have not yet been subjected to scientific validation *vis-à-vis* their quality, safety and efficacy.

In recognition of the crucial role African traditional medicine plays in health care delivery in Africa, in 2001, the Regional Strategy on Traditional Medicine for WHO African region was adopted. The Regional Strategy aims at promoting the development and institutionalization of traditional medicine into the public health care system. Furthermore, in 2001, the Summit of African Heads of States declared 2001-2010 the Decade for African Traditional Medicine.

In this paper, issues related to the development of traditional medicines with some examples will be discussed briefly.

2 DEVELOPMENT OF TRADITIONAL MEDICINES

In order to explore the therapeutic potentials of African Traditional Medicine, issues related to the Traditional Health Practitioner (THP) should be adequately addressed. One key element which can facilitate the cooperation of THP is recognition. The national health authorities need to utilize the various documents developed by the WHO for institutionalization of traditional medicine and recognize the THPs as holders of indigenous medical knowledge (IMK). For example, the generic document on National Policy on Traditional Medicine can be appropriately modified and adopted. The policy has provisions for the Traditional Health Practitioners Council, registration of THPs, and professional curriculum for THPs, among others. Subsequently, Traditional Medicine Law and applicable regulations can be developed. Furthermore, relevant government structures including directorate or department of Traditional Medicine should be established. These instruments and declaration in a Government gazette *vis à vis* official recognition of Traditional Medicine will encourage the THPs to collaborate with the Government and biomedical researchers. In such collaboration arrangements, an appropriate Memorandum of Understanding (MOU) or Memorandum of Agreement (MOA), which recognizes the THPs as partners in the product development process, owner of the IMK and consequently beneficiary of the commercial benefits of the IMK, should be developed. Similarly, the THPs will be trained in the proper use of the code of conduct by the National THP Council which, among others, has the powers to sanction any defaulting THP.

The immediate and urgent assignment is to document recipes held by genuine THPs using appropriate documentation tool prepared by WHO. The goal is to develop ethnomedical data which contains the best practices in African Traditional Medicine. A solid biodiversity policy which among others covers the cultivation and conservation of medicinal and aromatic plants as well as appropriate instruments for combating bio-piracy needs to be developed by various African countries. Such a step will promote the enforcement of Intellectual Property Protection policy.

One of the most crucial factors regarding the research and development of therapeutic potentials of African indigenous medical knowledge is intellectual property rights (IPR). This will require an appropriate Memorandum of Agreement to be developed which must include material transfer mechanism, equitable sharing of benefits, etc. The process of developing the MOU should be transparent, all inclusive and allow for development of mutual trust. Closely linked to IPR is the important issue of a benefit sharing arrangement that is acceptable to all the stakeholders. There should be both monetary and non-monetary components in the IPR document. Furthermore, the mechanisms for implementing the benefit sharing arrangement should be included in the IPR Memorandum of Agreement (for example, Trust Fund).

There are research institutions which have the primary mandates to undertake research and development of traditional medicines. Examples of such institutions include the Institute of Traditional Medicine (Mali), Centre for Medicinal Plant Research (Ghana), National Institute for Pharmaceutical Research and Development (NIPRD, Nigeria), Kenya Medical Research Institute (Kenya), and Council for Scientific and Industrial Research (South Africa).

After an ethnomedical survey undertaking by the University of Ougadougou, Burkina Faso, some THPs were identified who had been treating people living with HIV/AIDS. With the support of the WHO, a pilot clinical study was conducted using three different herbal medicines and a sample size of 1246 people living with HIV/AIDS (PLWAs). The study lasted 12 months. The results indicated that all the study participants had increased quality of life, CD4 values significantly increased in 75% of the participants, while 68% of the participants experienced profound decreases in viral load measurements. Furthermore, there was a simultaneous increase in CD4 and decrease in viral load of 63% of study participants.

In a similar study, PROMETRA (an NGO) in Senegal conducted pilot clinical trials with African herbal medicines in PLWAs. The data showed improved quality of life of patients, weight gain, reduced viral load and increased CD4 values.

The NIPRD developed a standardized phytomedicine for the treatment of malaria following the model it used to develop NIPRISAN for the management of sickle cell anaemia.[1] Non-clinical efficacy and safety profiles of the phytomedicine indicated that LD50 > 2000mg/kg per oral in rats and mice. Furthermore, there was no toxicity on the organs in a 28 days study and no abnormality in liver enzymes, while both creatinine and urea were normal. Similarly, triglycerides, cholesterol and glucose were not affected by the phytomedicine.

In a pilot clinical trial which was an open design conducted in two public hospitals in Nigeria, 29 volunteers with acute malaria infection were involved in the study. Both clinical and parasite counts were done. The result showed that all the volunteers responded to the phytomedicine with 100% parasite clearance and total remission of clinical signs associated with malaria infection. The volunteers on the control arm (16) were given fansidar. Only 40% of them responded to fansidar. The statistical analysis showed that the difference between the phytomedicine-treated and fansidar-treated groups was profound ($p < 0.0005$, paired t-test).

3 CONCLUSION

These few examples attest to the immense potentials of the therapeutic benefits hidden in Africa's biodiversity and IMK. In order to advance the development of standardized phytomedicines based on Africa's IMK, there are some strategic issues which need particular attention. The application of WHO Good Laboratory Practice (GLP) principles leading to generation of solid scientific data which can be published after patenting the new innovation is crucial. Ethical review systems need to be strengthened in health institutions in sub-Saharan Africa. The mandatory evaluation of study protocols and subsequent monitoring of clinical trials by national drug regulatory authorities are generally weak. The involvement of the private sector in research and development efforts by biomedical researchers is almost non-existent. It is vital to stimulate the public-private partnership mechanism and national governments are in the best position to do it. Similarly, investment by national government in health research is poor and inconsistent. The national governments need to formulate national Traditional Medicine and Biodiversity policies and enact appropriate laws. Furthermore, MOA *vis-à-vis* IPR should be cleared by appropriate national agencies to guarantee fair and equitable benefit sharing mechanisms.

Reference

1 C. Wambebe, Guest Editorial: *Innovation and Discovery*, 2006, **18**, 1.

NATURAL PRODUCTS FROM PLANT BIODIVERSITY AND THEIR USE IN THE TREATMENT OF NEGLECTED DISEASES

J.O. Midiwo

Department of Chemistry, University of Nairobi, P.O. Box 30197, Nairobi, Kenya.

1 INTRODUCTION

Neglected diseases are a group of infectious diseases that are endemic in the low-income populations in the developing world. They cause death to about 500,000 to 1m people every year, a death rate that is comparable to that of the HIV/AIDS pandemic. They are easily preventable but are a big problem in the affected areas. There are supposed to be about 13 distinct diseases but they can be classified as those caused by trypanosomal parasites, helminth parasites, bacteria and viruses.

Trypanosomal diseases include Kala-Azar or visceral leshmaniasis, which is a severe form of leshmaniasis, African sleeping sickness (African trypanosomiasis), which is experiencing resurgence, and Chaga's diseases (American trypanosomiasis), the current drugs for which are severely toxic, even though the disease does not kill as such.

Helminth (worm) parasite diseases include schistosomiasis, which is currently treated with praziquantel, an inexpensive drug but one that cannot stop re-infection, onchocerciasis (river blindness), on which anthelmintic treatments are being tried, drancunculiasis (guinea worm), which has an eradication target of 2009, and lymphatic filariasis, which causes elephantiasis and is currently being managed by anthelmintic treatments. The other helminth problems include soil-transmitted worms such as ascariasis (round worms) where efficient drugs exist but control involves mainly observance of hygiene, and trichuriasis (whipworm) and hookworm for which there are also good control methods.[1]

Bacterial infections are the other common neglected diseases. They comprise leprosy and trachoma, which are controlled by antibiotic treatments, Buruli ulcer, managed by surgery and antibiotics, cholera, caused by *Vibrio cholerae* living in contaminated drinking water, and for which antibacterials and isotonic fluid replacement therapy are recommended.

Viral infections are yellow and dengue fevers, caused by flaviviruses transmitted by *Aedes aegypti*, and Japanese encephalitis, also caused by a flavivirus, which is spread by *Culex tritaeniorhynchus*.[1] All these are controlled by vaccine administration.

However, the WHO Innovative and Intensified Disease Management (IDM) group considers only the following as neglected tropical diseases (NTDs): Buruli ulcer, Chaga's disease, cholera, human African tryapanosomiasis and leshmaniasis. They consider them as neglected relatives of the so called big three - HIV/AIDS, tuberculosis and malaria, which receive regular research funding "voice".

Many organizations do not agree with this definition. For example, the Drugs for Neglected Diseases Initiative (DNDi) organization considers the following as the neglected diseases: African tryapanosomiasis, leshmaniasis, Chaga's disease and malaria, thus leaving out Buruli ulcer but including malaria. DNDi believes that malaria does not get enough attention amongst researchers relative to its perilous effects on populations in tropical areas. Their opinion is that "neglected diseases" should be "diseases that do not get enough funding relative to their impact on the population".[2]

Clearly there is African folklore (mostly medicinal herbs) about the management of neglected diseases. It would be prudent to explore the potential of certain local herbs for the development of phytomedicines for local populations for neglected diseases. Such practices have even been recorded in monographs such as Medicinal Plants of East Africa (MPEA),[3] which is quite useful. MPEA lists plants traditionally in the region for certain disease categories - helminthiasis, fever including malaria, and bacterial infections, amongst others. It describes synoptically the methodology of preparing the herbs from the plants and the mode of usage.

2 HELMINTHIASIS

Helminthiasis as a disease group actually covers many of the neglected diseases. More than five hundred East African plants are listed by Kokwaro[3] in MPEA as emetics/ purgatives, alleviators of infections of intestinal worms, such as for hookworm, roundworm, tapeworm and threadworm, general anthelmintics, and treatments for bilharzia and filarial infections. Some plants overlap amongst the categories of helminthiasis by different ethnic groups and should be investigated for their high potential effectiveness in modern medicine. A few such plants have been studied in our laboratory. They are the Myrsinaceae,[4] which are mentioned in several categories of anthelminthiasis, and *Rumex*[4] species, listed as laxatives which can also be used expel helminths from bowels. The usage and effects of Myrsinaceae herbs are indicated in Table 1.

Table 1 *Traditional uses of the Myrsinaceae*

PLANT	USE
M. africana	2-3 handfuls of fruits are chewed for round worm and tapeworm treatment and as remedy for chest pains and stiff joints. No harm is done when too much is taken.
E. schimperi	Fruits chewed as both vermifuge and purgative. Dried fruits and roots are boiled or soaked in water and infusion drunk for intestinal worms.
R. melanphloes	Fruits used as anthelmintic when chewed and eaten in porridge to expel intestinal worms.
M. lanceolata	Fruits used as purgative to remove worms and as remedy for sore throats or eaten to cure tapeworm

We have looked at the phytochemistry of the Kenyan Myrsinaceae and we find that it corresponds with that already reported in the literature. They are harbingers of long alkyl side chain 2,5-dihydroxybenzoquinones.[4] We managed to group them chemotaxonomically into two sub-families, the Myrsinodae and the Maesodae, using these markers, which coincide with their morphological classification. The Myrsinodae, which includes the species *Myrsine africana, Rapanea melanphloes Embelia schimperi* and *E. keniensis*, are chemically typified by the existence of embelin and rapanone, while the Maesodae, represented monotypically by *Maesa lanceolata* in Kenya, is typified by the existence of maesaquinone. We have isolated several other benzoquinones from Kenyan Myrsinaceae and have reported them in the literature.[4]

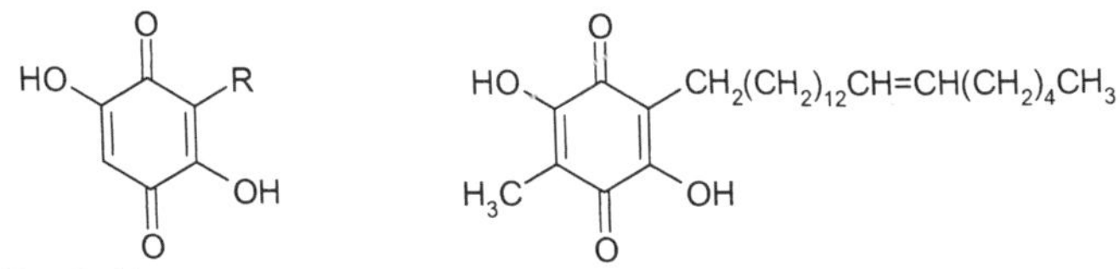

$R = C_{11}H_{23}$ -- Embelin

$R = C_{13}H_{27}$ -- Rapanone

Maesaquinone

Figure 1 *Typical benzoquinones in the Myrsinaceae*

Rumex species are widely used locally for the control of intestinal helminthic conditions. There are five Rumex species in Kenya - *Rumex abyssinicus, R. usambarense, R. bequaertii, R. ruwenzoriensis* and *R. crispus*, according to their grouping in the "Key to Species". The compounds that have significant distribution are the common anthraquinones emodin, physcion and chrysophanol, along with the polyketide naphthalenic compound, nepodin, whose distribution is in accordance with the Key.[4] The compounds have been reported to have a purgative effect; no doubt this is the mechanism by which they exert their traditionally attractive anthelmintic activity. The traditional applications of some of the species are shown in Table 2.

Table 2 *Traditional applications of some* Rumex *species*

SPECIES	TRADITIONAL USES
R. abyssinicus	Leaves and stems are powdered and sap used as treatment for pneumonia and cough. Roots powdered and applied to wounds or infusion drunk as remedy for stomach-ache.
R. bequaertii	Roots used for treatment of abscesses, or abdominal pains.
R. usambarense	Leaves used for coughs or ground and mixed with water to stop stomach-ache. Whole plant is used for treatment of small pox.

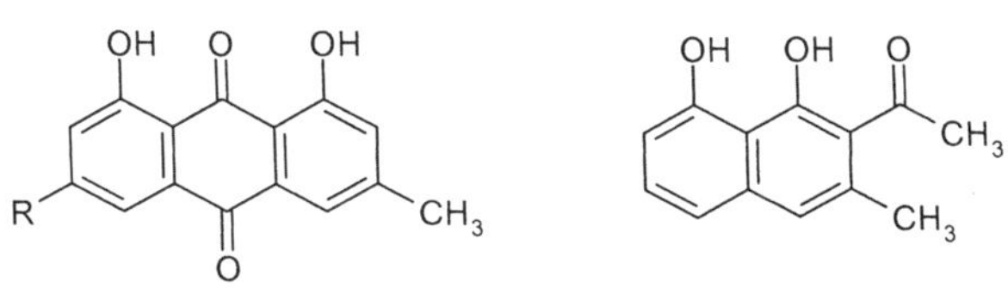

R=OH ------ Emodin

$R=OCH_3$ -- Physcion

R=H -------- Chrysophanol

Nepodin

Figure 2 *Pigments in* Rumex *species of Kenya*

Considering the high concentrations of bioactive principles in these Myrsinaceae, *Rumex* species and similar anthelmintic plants, it is suggested that they could be pursued for formulation of cheap phytomedicines for use by the local populations. Some of the herbs, like the berries of the Myrsinaceae, are usually added into food as medicinal spices for the desired effects for intestinal worm expulsion, and they can be developed as food supplements to control intestinal worm proliferation.

3 MALARIA

Malaria is a serious disease whose progression may lead to death. Its symptoms of high fever, chills, weakened joints, and flu-like illness are however well recognised by people in malaria endemic areas such as Kenya. Illness and death from malaria are largely preventable. Malaria is caused by *Plasmodium* parasites which are transmitted by mosquito vector. The most common parasite in Kenya, which is also the most virulent, is *Plasmodium falciparum*; the others are *P. vivax, P. ovale* and *P. malariae.* The main mosquito vector in Africa and which, unfortunately, is also the most efficient is *Anopheles gambiae.*

There are 300-500 million malaria infections world-wide every year. About 80% of these are in Africa, leading to 1.75-2 m deaths annually, mostly of children under five years old. Worldwide, 3000 children die every day from malaria. This is an equivalent of seven jumbo jets full of children crashing on a daily basis! Approximately 90% of malaria deaths are in Africa, and malaria constitutes 10% of Africa's total disease burden, 40% of its health expenditure and 30-50% of its in-patient cases.[5] The total African cost estimate is in the range of US$12b annually. In Kenya, 22 million people are at risk, of which 70% are in rural areas. About 34,000 Kenyan children die every year from malaria, compared to a total estimate of 42,000 people dead.[6]

3.1 Malaria control methods

Malaria can be controlled using three approaches - drug therapy, vector eradication and the use of vaccines (in the future).

Many anti-plasmodial drugs have been developed during the course of time. Quinine was identified as a useful drug from the South American plants of the genus *Cinchona* since 1645. The extracts of the species were used for a long time without cases of parasite resistance being observed. Due to the fact that quinine exhibited relatively high mammalian toxicity, and the issue of ready availability, its synthetic analogues were sought. Therefore drugs like chloroquin and mefloquin were developed; they had the feature of being cheap, readily available and had much lower toxicity compared to the prototype quinine. On the other hand, due to wide usage, these synthetics soon became ineffective because the parasite in many parts of the world developed resistance to them.

Table 3 shows the number of years it took before some preferred anti-malarial drugs experienced resistance from parasites. Clearly, there is a difference between the natural drug quinine and the synthetics in that it took much longer for quinine to face this fate. The current first line drugs, however, have fickle longevity. The problem of resistance development makes the search for anti-malarial drugs an elusive and continuous undertaking. Indeed, due to this problem, the situation of malaria therapy is worse now than it was 20 years ago. In Kenya the use of chloroquin has been stopped for some time. According to the Kenyan Director of Medical Services, the government

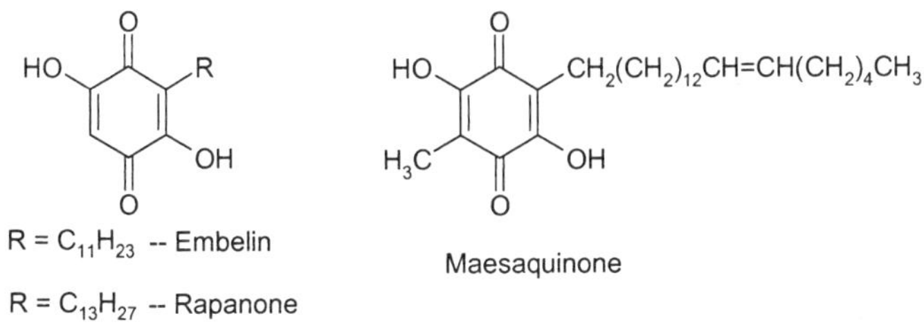

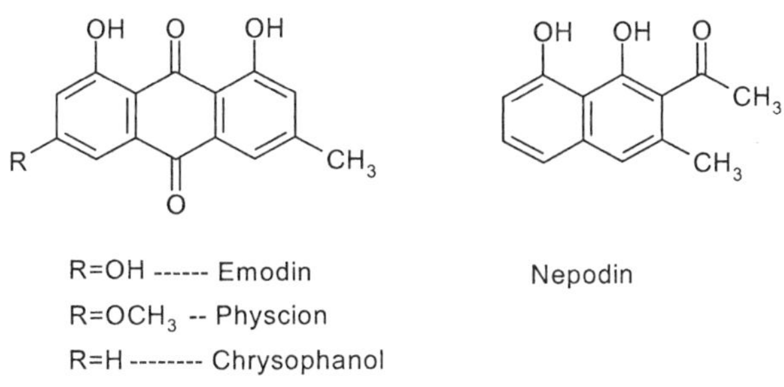

is to discourage the use of fansidar and metakalfin, both of which are sulfadoxin/pyrimethamine combinations, yet they have been the recommended choice for less than three years.

Table 3 *Time taken by various anti-malarial drugs to experience resistance from the parasite.*

Anti-malarial drug	Introduced	1st resistance reported	Difference
Quinine	1645	1910	278
Chloroquine	1945	1957	12
Proguanil	1948	1949	1
Sulfadoxin/Pyrimethamine	1967	1967	0
Mefloquine	1977	1982	5
Atovaquone	1996	1996	0

The other natural compound and its derivatives that are now being widely constituted into anti-malarial drugs is artemisinin. This compound was isolated from the age-old Chinese anti-malarial drug, *qinq hao su*, isolated from the plant *qinq hao* or *Artemisia annua*, which is now a widely grown crop in Kenya and East Africa in general. Artemisinin is an excellent anti-plasmodial whose derivatives are now being constituted as artemisinin combination therapy drugs (ACTs) by different companies. An example of an ACT is Coartem, a combination of artemether and lumefantrine, from Novartis.

ACT drugs are quite expensive compared to single compound drugs like chloroquin and mefloquin or the sulfadoxin/pyrimethamine combinations. The Global Fund for malaria and AIDS has allocated funds to purchase and provide Kenya with free ACTs for two years to help in the war against malaria. It is going to be very difficult for Kenya to sustain availability of these drugs after the GF period runs out. The majority of Kenyans are at present not accessing proper anti-malarial drugs and this is expected to be the case after the GF period as well; perhaps another approach is necessary.

A decoction of *Artemisia annua*, using 5 g of dry leaves and one liter of water and taken in four parts in one day, every day, for five days, completely gets rid of malaria parasitemia. This method compares to that primed by the Chinese over millennia without any observation of resistance. The most likely reason for the lack of resistance is because the decoction is a combination of many compounds - 29 sesquiterpenes like artemisinin and 36 flavonoids acting together (synergism). It is in fact known that when artemisinin is removed from the mixture the remaining combination is still reasonably anti-plasmodial.[7] A logical conclusion is that to get adequate medication to the villages and, indeed, even to the urban areas, instructions to use anti-malarial herbs like *A. annua* need to be propagated. We have also identified other plants from our folk lore, whose extracts and compounds we have determined to have anti-plasmodial activity, which can also be used.

3.2 *Erythrina abysinica* and *Polygonum senegalense*: examples of sources of potential Kenyan plant anti-malarials.

Erythrina abysinica is a Papilionaceae plant whose stem and root barks have been used against malaria fever by many communities in Kenya.[3] Chimpanzees at Mahale Game Reserve in Tanzania are also known to partake of its bark when they have fever-like symptoms.[8] Crude ethyl acetate extracts of the stem and root bark were subjected to tritium hypoxanthine uptake experiments to determine their anti-plasmodial activity.[9] They were tested against two strains of *Plasmodium falciparum,* Sierra Leone I chloroquine-sensitive (D6) and Indochina-chloroquin-resistant (W2). It was observed that the extracts of the stem bark had LC_{50} values of 7.9 and 0.49 μg/ml against the sensitive and resistant strains, respectively, while the extracts of the root bark had LC_{50} values of 5.3 and 0.64 μg/ml, respectively, against the same two strains. These are good values which approach those of chloroquin and quinine at about the nanogram/ml level and explain the preference for this herb in traditional malaria therapy.

The mixture of compounds contained in the crude extract were separated by chromatography on silica gel using various organic solvents. This process led to the isolation of three chalcones, 11 prenylated flavanones, three pterocarpanoids and one isoflav-3-ene (Figure 3). While the details of the characterization of these compounds have been reported elsewhere,[10] it is worth pointing out that in the bioassay assessment referred to above, they had reasonable LC_{50} values (Table 4). They had antimalarial activities comparable to those of the crude extracts and should be the principles responsible for the activity.

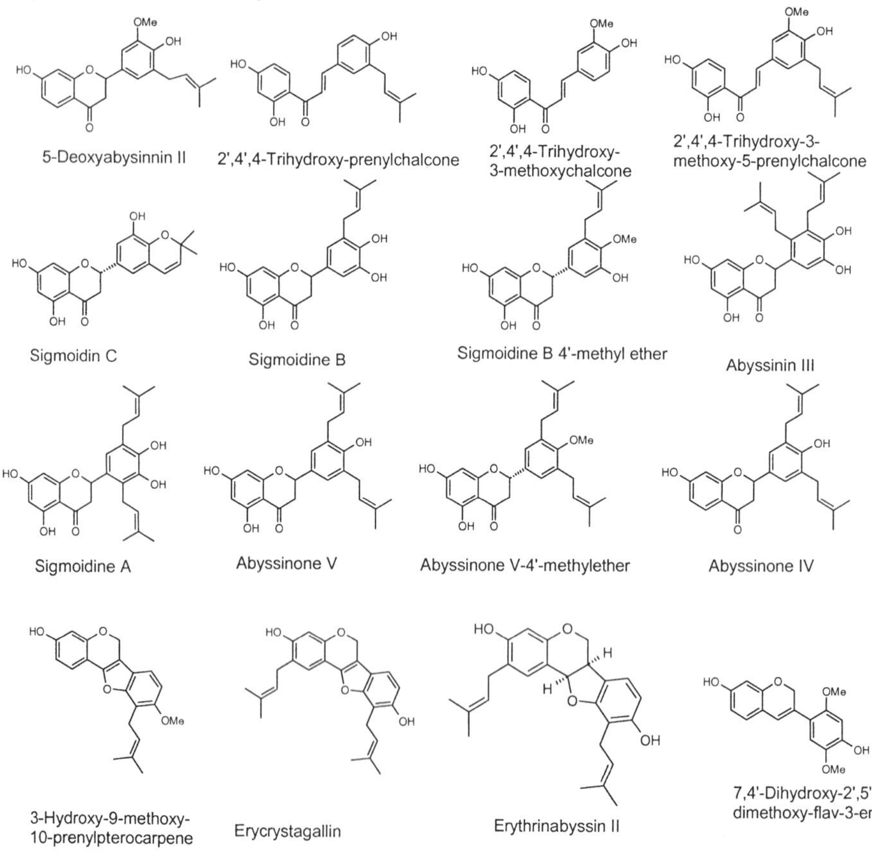

Figure 3 *Flavonoids from* Erythrina abyssinica *bark.*

Table 4 In vitro IC_{50} *values of flavonoids against W2 and D6 strains of* P. falciparum.

Flavonoids	IC_{50} (μM)	
	D6	W2
Chalcones		
5-Prenylbutein (1)	10.3 ± 1.3	11.2 ± 1.9
Homobutein	15.0 ± 2.8	16.1 ± 2.1
Licoagrochalcone A	12.7 ± 3.2	12.0 ± 2.6
Flavanones		
5-Deoxyabyssinin II (2)	13.6 ± 0.9	13.3 ± 1.5
Abyssinin III	5.8 ± 1.1	5.2 ± 1.7
Abyssinone IV	5.4 ± 1.5	5.9 ± 1.8
Abyssinone V	4.9 ± 0.8	6.1 ± 1.3
Abyssinone V - 4'-methyl ether	11.3 ± 2.1	11.1 ± 2.4
Sigmoidin A	5.8 ± 0.6	5.9 ± 1.1
Sigmoidin B	8.1 ± 2.2	9.3 ± 2.7
Sigmoidin B - 4'-methyl ether	13.0 ± 2.0	12.7 ± 2.9
Sigmoidin C	17.8 ± 3.6	15.8 ± 3.9
Sigmoidin E	9.1 ± 2.3	11.8 ± 2.5
Isoflavonoids		
3-Hydroxy-9-methoxy-10-prenylpterocarpene	18.2 ± 3.6	20.2 ± 3.2
7,4'-Dihydroxy-2',5'-dimethoxyisoflav-3-ene	22.0 ± 2.4	24.9 ± 3.0
Reference drugs		
Chloroquin	0.009 ± 0.002	0.008 ± 0.003
Quinine	0.04 ± 0.01	0.21 ± 0.01

Polygonum senegalense is a Polygonaceae weed whose leaves have been used traditionally for treatment of wounds and removal of ectoparasites from cattle. However, there are no records of its use as an anti-malarial. It is an interesting weed whose aerial parts (stems, leaves and flowers) are covered by a viscous exudate and whose leaves deposit a yellow smear when rubbed. The exudate apparently has ecological functions, acting as a sun screen and having anti-desiccant, and anti-fungal effects.[11] The exudates (which are up to 17% of the dry leaf weight) are composed of chalcones, flavanones, a flavone and a homisoflavanone (Figure 4) which we have reported on severally along with their mosquito larvicidal and locust anti-feedant activities.[4,15]

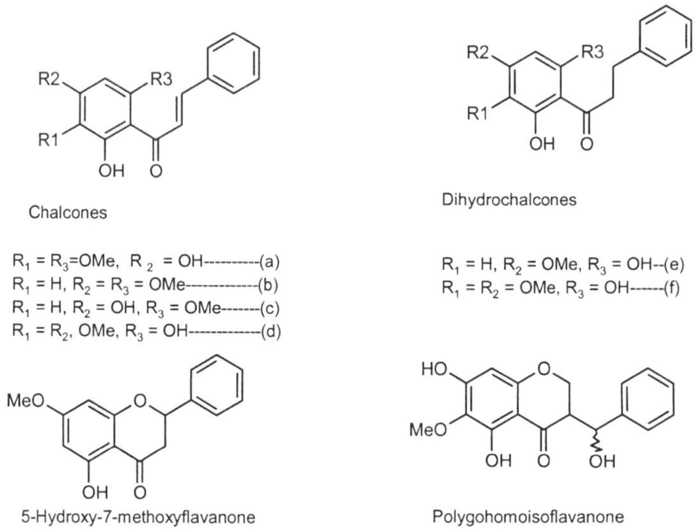

Figure 4 *Surface exudate compounds of* Polygonum senegalense.

The apparent bioactivity of the chalcones in the mixture, coupled with reports that chalcones are known to have anti-plasmodial activity,[14] and that they are available in large quantities, led us to submit them to assays for anti-plasmodial activity; the chalcones, flavanones and the homoisoflavanone (Table 5) were found to have reduced but still significant anti-plasmodial activities with IC_{50} values in the range of 2-20 μg/ml. This is an illustration of the principle that plant extracts and component compounds that are bioactive should be surveyed for anti-malarial activity and for activities against other diseases for which they are not already known.

Table 5 Anti-plasmodial activities of *P. senegalense* surface exudate compounds against the D6 and W2 strains of *Plasmodium falciparum*

Compounds	(IC_{50} in μM)	
	D6	W2
Polygohomoisoflavanone	19.2 ± 0.6	18.2 ± 1.8
a Chalcone	16.6 ± 2.3	17.8 ± 0.08
b	15.7 ± 1.2	11.3 ± 0.2
c	14.0 ± 1.7	9.5 ± 1.1
d	3.1 ± 0.8	2.4± 0.3
e	23.5 ± 0.8	22.8 ± 1.8
f	11.8 ± 0.3	12.9 ± 0.5
5-Hydroxy-7-methoxyflavanone	16.3 ± 1.8	21.8 ± 1.2
Reference drugs		
Chloroquine	0.008 ±0.002	0.075 ± 0.002
Quinine	0.050 ± 0.02	0. 28 ± 0.02

4 CONCLUSION

Medicinal plants are useful sources of anti-plasmodial compounds that can be packaged as new anti-malarial drugs. It is worth pursuing the concept that mixtures of such compounds from plants can be formulated together for synergistic activity, in which case the effective concentrations can be significantly reduced. It is the belief that this "shift in paradigm" from single or small number compound therapies may in the future lead to the development of more efficient and cheaper drugs which also circumvent resistance development.

References

1 WHO Publication, January 2008.
2 Drugs for Neglected Diseases initiative (DNDi), Wikipedia 2003.
3 J.O. Kokwaro, Medicinal Plants of East Africa, 2nd edition. Kenya Literature Bureau 1993.
4 J.O. Midiwo, A. Yenesew, B.F. Juma, S. Derese, J.A. Ayoo, A.O. Aluoch and S. Guchu, Bioactive compounds from some Kenyan ethnomedicinal plants: Myrsinaceae, Polygonaceae and *Psiadia punctulata*. Phytochemical Reviews, 2002, 1: 311-323.
5 The World Health Report 2001 pp 144-145, 150-151.
6 DMS Kenya 2006. Figures provided by Director of Medical Services of Kenya as reported in the local press.
7 G. François, M. N'Guessan, M. Koffi, C. Dochez, M. Wéry and L. Triest, *Planta Medica,* 1993, **59**, A677.
8 Huffman, M. A., Ohigashi, H., Kawanaka, M., Page, J. E., Kirby, G.C., Gasquet, M., Koshimizu, K. African great ape self medication: A new paradigm for treating parasite diseases with natural medicines. In: "Towards natural medicine research in the 21st century, Ebizuka, Y. (ed). Elsevier Science B.V. Excerpta Medica, Amsterdam, 1998 pp 113-123.
9 A. Yenesew, S. Derese, J.O. Midiwo, H.A. Oketch-Rabah, J. Lisgarten, R. Palmer, M. Heydenreich, M.G. Peter, H. Akala, J. Wangui, P. Liyala, and N.C. Waters, *Phytochemistry*, 2003, **64**, 773.
10 A. Yenesew, M. Induli, S. Derese, J.O. Midiwo, M. Heydenreich, M.G. Peter, H. Akala, J. Wangui, P. Liyala and N.C. Waters, *Phytochemistry*, 2004, **65**, 3029.
11 E. Wollenweber, 1985. Flavonoid aglycones as leaf exudate constituent in higher plants, pp 155-169. In: Flavonoids and Bioflavonoids (eds. L. Farkas, M. Garbor, and F. Kelley). Institute of Botany, university of Darmstadt D610U, Darmstadt, FRG.
12 J.O. Midiwo, J.J. Matasi, O.M. Wanjau, R.W. Mwangi, P.G. Waterman and E. Wollenweber, *Bull. Chem. Soc. Ethiop.,* 1990, **4**, 123.
13 J.O. Midiwo, N.K. Gikonyo, D.O. Wanjau, J.J. Matasi and P.G. Waterman, *Bull. Chem. Ethiop.*, 1992, **6**, 119.
14 M. Liu, P. Wilrat, M.L. Go, *J. Med. Chem.*, 2001, **44**, 4443.
15 J.O. Midiwo, F.M. Omoto, A. Yenesew, H.M. Akala, J. Wangui, P. Liyala, C. Wasunna and N.C. Waters, *Arkivoc*, 2007, (ix), 21.

IN VITRO AND *IN VIVO* ANTIPLASMODIAL ACTIVITY OF KENYAN MEDICINAL PLANTS

M.O. Nanyingi,[1,2] K.B. Kipsengeret,[1,2,3] C.G. Wagate,[2] B.K. Langat,[4] L.L. Asaava[2] and J.O. Midiwo[5]

[1]Kenya Scientific Analysts, P.O. Box 531-00202, Kenyatta National Hospital, Nairobi, Kenya, Email: mnanyingi@gmail.com.
[2]Department of Public Health, Pharmacology and Toxicology, Faculty of Veterinary Medicine, University of Nairobi, P.O. Box 29053-00625, Nairobi, Kenya.
[3]Center for Public Health Research, Kenya Medical Research Institute, P.O. Box 20752-00202, Nairobi, Kenya.
[4]Centre for Biotechnology Research and Development, Kenya Medical Research Institute, P.O. Box 54840-00200, Nairobi, Kenya.
[5]Department of Chemistry, University of Nairobi, P.O. Box 30197, Nairobi, Kenya.

1 INTRODUCTION

Malaria is one of the most prevalent diseases in the world and a major public health problem, affecting 40% of the world population. It infects about half a billion people causing about 3 million deaths a year with 90 % of them being in sub-Saharan Africa.[1,2] In Kenya, malaria is the leading cause of morbidity and mortality, accounting for 30-50% of all outpatient attendance and 20% of all admissions to health facilities. The infection commences with the intravenous inoculation of sporozoites by infected mosquitoes. These invade hepatocytes and undergo multiplication before merozoites rupture these cells to invade erythrocytes and commence the erythrocytic cycle. The pathogenic process occurs only during the erythrocytic cycle. During this stage, there is a huge, periodic amplification of the size of parasite populations that may enhance the probability of differentiation to gametocytes, the stage infectious to mosquitoes. A peculiarity of *P. falciparum* is its ability to adhere to venular endothelium (cytoadherence) of erythrocytes infected with maturing parasites.

Kenya has revised its treatment policy by adopting artemisinin combination therapy (ACT) as the first line drugs for treatment of uncomplicated malaria. Spread of multidrug-resistant (MDR) strains of *Plasmodium* and the adverse side effects of the existing antimalarial drugs have necessitated the search for novel, well tolerated and more efficient antimalarial drugs.[3] Indigenous plants are important sources of biologically active compounds and have potential for the development of novel antimalarial drugs.[4, 5, 6]

The increasing prevalence of strains of *P. falciparum* that are resistant to chloroquine (CQ), a blood schizontocide which had been efficacious, safe, accessible and affordable, poses a serious problem for malaria control, predisposing Africa to an unprecedented situation since the only affordable treatment options are rapidly losing therapeutic efficacy.[7] Drug-resistant strains of *P. falciparum* are endemic in many areas of the world and the majority of conventional antimalarial drugs have been associated with treatment failure. These developments and the difficulty of creating efficient vaccines, coupled with adverse reactions to chemotherapy, underline the urgent need for novel, cheap, safe and efficacious antimalarial drugs.[8] It is estimated that 80% of people worldwide use herbal remedies, due to limited access to modern medicine because of low income and the shortage of efficient health care facilities.[9] There is a dearth of evidence on the efficacy and safety of these remedies, despite the fact that validation of traditional practices could lead to innovative strategies in malaria control.[10]

Synthetic antimalarials developed from herbals include chloroquine, primaquine, proguanil, pyrimethamine and mefloquine. Botanicals represent a diverse arsenal of molecules that could constitute lead compounds for new antimalarial drugs, such as artemisinin, isolated from *Artemisia annua*.[11] Several studies have been undertaken to evaluate the inhibitory effects of various plants extracts on *P. falciparum* in culture.[12,13] The *in vivo* antiplasmodial effects of several plant extracts have been studied on *Plasmodium berghei*[14] and *P. yoelii.*[15] The majority of the plants that we screened for antimalarial activities had similar ethnopharmacological use among different Kenyan ethnic groups.[16, 17]

Based on this, the current study evaluates the *in vitro* antiplasmodial activities and *in vivo* antimalarial properties of six plants used in Samburu traditional medicine in Kenya, viz. *Azadirachta indica* A. Juss (Mimosaceae) MN 29, *Warburgia ugandensis* Sprague (Canellaceae) MN 37, *Myrsine africana* L (Myrsinaceae) MN 7, *Maytenus senegalensis* (Lam.) (Celastraceae) MN 47, *Teclea simplicifolia* (Engl.) (Rutaceae) MN 42, and *Harrisonia abyssinica* Oliv (Simaroubaceae) MN 24. The cytotoxic activities and acute toxicity of these plants were determined using Vero E6 cells.[18, 19]

2 METHODS

2.1 Plant materials and extraction

The six plant samples used in this study were collected between May and August 2007 from the Samburu district of Kenya based on a knowledge of ethnopharmacological use established through participatory epidemiological approaches (questionnaires, interviews, focus group discussions and transect walks) with local communities and traditional health practitioners (THP). Ethnomedical data included vernacular names (in parentheses) and the parts used in preparation of the herbal antimalarial remedies: *A indica* (Mwarubaini) stem bark/leaves, *W. ugandensis* (Sokorioi) stem bark/leaves, *M. africana* L (Seketet) seeds, *M. senegalensis* (Laimurunyai) leaves, *T. simplicifolia* (Lgelai) roots and *H. abyssinica* (Lasaramai) roots/leaves. The plants were identified by a taxonomist at the herbarium of the Department of Land Resource Management and Agricultural Technology (LARMAT), University of Nairobi, where voucher specimens were deposited. The plant samples were then catalogued, chopped into small pieces, air-dried at room temperature (25 ◦C) under shade, and ground into powder using an electric mill (Arthur H Thomas, USA). Methanol extracts (50 g powder in 500 ml of solvent) were prepared by maceration of the plant material with methanol at room temperature for 48 h. The mixture was filtered and the filtrate concentrated to dryness *in vacuo*. The yields of the methanol extracts ranged between 2.0 and 6.5% of the dry weight of the plant. The dry solid extracts were stored at −20 ◦C in airtight containers and lyophilized extracts were kept at +4°C until used.

2.2 Preparation of test extracts

Stock solutions of crude extracts (250 μg/ml) were made with sterile water (deionized and autoclaved) and consecutively filtered through 0.45 and 0.22 μm microfilters in a laminar flow hood. The water-insoluble extracts were first dissolved in DMSO (solvent concentration <0.02%) (Sigma Chemical Co., St. Louis, MO, USA). The reference drugs CQ diphosphate and artemisinin, each at a concentration of 1 μg/ml, were similarly prepared and all solutions were stored at −20 ◦C and retrieved only when used.

2.3 Parasite specimens and culture

To test for *in vivo* antiplasmodial activities of the methanol extracts, the mouse-infective, the CQ-susceptible (D6) and resistant (W2) clones donated by the Division of Experimental Therapeutics, Walter Reed Army Institute of Research, Washington DC, were used. To test the *in vitro* antiplasmodial activities of the plant extracts, the CQ-sensitive *P. berghei* (strain ANKA), donated by International Livestock Research Institute (ILRI), Kenya was used.

P. falciparum culture of D6 (CQ-sensitive isolate from Sierra Leone) and W2 (CQ-resistant isolate from Indochina) were used in the study. The culture medium was a variation of that described by Trager and Jensen.[20] Parasites were cultivated in 75 cm^3 flasks containing RPMI 1640 medium (20ml) supplemented with 10% human serum, 25 mM *N*-2-hydroxyethylpiperazine-*N*-2-ethanesulfonic acid (HEPES) and 25mM $NaHCO_3$. Parasitaemia was maintained between 1 and 6%. Uninfected human blood group O+ erythrocytes (<28 days old) served as host cells. The cultures were incubated at 37°C in an atmosphere of 3% CO_2, 5% O_2 and 92% N_2 (EAO®, Nairobi) with 90% humidity. Renewal of the medium and microscopic observation (×100) by blood smears fixed with methanol and stained with 10% Giemsa's stain were performed on a daily basis.[21]

2.4 Determination of parasitemia

Parasite growth was assessed by flow cytometry according to a methodology previously described using hydroethidine (HE, Interchim, Montluçon, France) that is converted by metabolically-active cells into ethidium salts.[22] After incubation with hydroethidine, parasitized and uninfected erythrocytes were all identified on the basis of fluorescence intensity and size. A flow cytometer FACSort (Beckton Dickinson, Paris, France) equipped with an argon laser (power 15mW and wavelength 488 nm) was used for parasitaemia reading.

2.5 In vitro antiplasmodial bioassay

The *in vitro* semi-automated microdilution assay technique that measures the ability of the extracts to inhibit the incorporation of [G-^{3}H] hypoxanthine (Amersham International, Buckinghamshire, UK) into the malaria parasite was used. The extracts were tested in duplicate at ten concentrations in two-fold dilutions and the experiment was repeated twice for each extract. Computation of the concentration of drug causing 50% inhibition of [G-^{3}H] hypoxanthine uptake (IC_{50}) was carried out by interpolation after logarithmic transformation of both concentration and counts per minute (cpm) values using the formula:

$$IC_{50} = \frac{\text{antilog}\,(\log X1 + [(\log Y50 - \log Y1)\,(\log X2 - \log X1)]}{(\log Y2 - \log Y1)}$$

Where $Y50$ was the cpm value midway between parasitisized and non-parasitisized control cultures and $X1$, $Y1$, $X2$, and $Y2$ were the concentrations and cpm values for the data points above and below the cpm midpoints.[23]

2.6 Determination of in vivo antimalarial activity

A previously described 4-day suppressive protocol was used to screen the extracts.[24] The chloroquine-sensitive *P. berghei* strain ANKA was used to assess the intrinsic *in vivo* antimalarial activity. The parasite strain was maintained by serial passage of blood from an infected mouse to a naive mouse. Female Swiss mice (6–7 weeks old; 20–22 g) were randomly infected by intraperitoneal (i.p.) inoculation of 107 erythrocytes parasitized with *P. berghei* in a saline suspension of 0.2 ml on day zero (D0) and allocated to several groups of five mice in each cage. They were fed on standard pellets and water *ad libitum.* The animals were housed in the small animal house at the International Livestock Research Institute. The Institute's Animal Care and Use Committee gave approval for the study.

Plant extracts were solubilized in 10% tween 80 (methanol extracts) or in saline (water extracts) and administered intraperitoneally once daily (D0 to D3) at a concentration of 100 mg/kg/day in a dose volume of 0.2 ml. Two groups (five mice each) served as negative and positive controls, respectively. The negative group received saline/tween 80 while the positive group was treated with the reference drug CQ diphospate at a dose of 5 mg/kg/day. Intraperitoneal *in vivo* antimalarial activity of the test drugs was assessed by monitoring mouse survival and parasitaemia, over that period. Each day from D1 to D4, thin blood smears were made from the tail of each mouse, stained with 10% Giemsa in phosphate buffer at pH 7.2, and examined microscopically for assessment of parasitaemia. The mean parasitaemia in each group of mice on day 4 was used to calculate the percentage chemo-suppression for each extract using the formula:

$$\frac{(A - B)}{A \times 100}$$

Where A was the mean parasitaemia in the negative control and B was the parasitaemia in the test group. Extract activity was determined by percentage reduction of parasitaemia in treated groups compared with untreated infected mice. For all the groups of experimental mice used, survival time in days was recorded and the mean for each group calculated.[25]

2.7 Acute toxicity

Healthy Swiss female mice weighing 15–25 g were divided into groups of five in each cage and had access to tap water and food, except for a short fasting period (12 h) before oral administration of a single dose of the extract. The extracts were dissolved in distilled water and administered by gavage at logarithmic doses ranging between 500 and 5000 mg/kg body weight to give five dose levels of 500, 889, 1581, 2812 and 5000 mg/kg body weight. The mice were monitored for sings of toxicity for 1 h post treatment and then intermittently for 4 h, and thereafter over a period of 24 h The mice were further observed for up to two weeks following treatment for any signs of toxicity, and the latency of death.[26]

2.8 Cell cytotoxicity assay

The cytotoxic concentration causing 50% cell lysis and death (CC_{50}) was determined for the extracts where kidney cells of African green monkey (Vero E6) were seeded at a concentration of 2.5×10^4 cells/well in 24-well plates and grown in minimum essential medium (MEM) at 37°C under 5% CO_2 for 48 h. The culture medium (MEM) was replaced by fresh medium containing an extract at various concentrations, and the cells further grown for 24 h. The cells were treated with trypsin and the number of viable cells determined by the trypan blue exclusion method. The concentration of herbal extract reducing cell viability by 50%, the CC_{50}, was determined from a curve relating percent cell viability to the concentration of the extract.

2.9 Statistical analysis

Percentage suppression of parasitaemia for the plant extracts was calculated as: 100–[(mean parasitaemia treated/mean parasitaemia control) × 100]. For comparison of average parasitaemia, one-way ANOVA and two-tailed Student's *t*-test were used with $P < 0.05$ being considered significant. The concentration of the plant extract required to induce a 50% decrease of infected erythrocytes (IC_{50} *Plasmodium*) was calculated by non-linear regression analysis processed on dose–response curves using the Table Curve Program.[27,28]

3 RESULTS AND DISCUSSION

The *in vitro* activities (IC_{50}) of the extracts against CQ-sensitive (D6) and resistant (W2) *P. falciparum* clones, and their cytotoxicities (CC_{50}) on mammalian Vero E6 cells, are shown in Table 1. The Selectivity Index (SI) of each extract, defined as the ratio of the CC_{50} value on the mammalian cells to that of the IC_{50} value on the CQ-sensitive *P. falciparum*, is also shown. Extracts with a high selectivity for the parasites offer the potential for safer therapy. The activity was categorized as high, moderate, and weak, respectively, for IC_{50} values of <10μg/ml, between 10 and 20μg/ml, and between 20 and 100μg/ml. Extracts with IC_{50} values >100μg/ml were categorized as inactive.

The two clones were highly susceptible (IC_{50} <8μg/ml) to the methanol extracts of *M. senegalensis* and *W. ugandensis*. The methanol extracts of *M. senegalensis* were the most active, with both extracts displaying high activity (IC_{50} <3μg/ml) for both clones. The CC_{50} value for the methanol extract was much higher than the corresponding IC_{50} value, indicative of high selectivity for the *Plasmodium* spp. Indeed, all cytotoxicity values of the extracts on Vero cells were higher than the corresponding IC_{50} values on *P. falciparum*. This was consistent with the *in vivo* safety of the extracts, where no acute toxicity was observed in mice (LD_{50} >5000 mg/kg.).

There was no correlation between the *in vivo* and *in vitro* antimalarial activity. This may be due to pharmacodynamics/pharmacokinetics (biotransformation and bioavailability) of the constituents and the active compounds *in vivo*. However, both extracts had similar survival times (14.0 and 15.0 days) to that of the reference drug CQ ($p > 0.05$).

The leaves of *Maytenus senegalensis* are used in East Africa for hookworm infestations. In the Sudan, the antiplasmodial activity of *M. senegalensis* leaves and stem bark was reported (IC_{50} 5.1 and 3.9μg/ml respectively).[29] Preliminary phytochemical analysis of the dichloromethane fraction of *M. senegalensis* revealed terpenoids and traces of phenolic principles. The antiplasmodial activity of *Maytenus* spp. is promising. The survival time was similar to that of CQ ($p > 0.05$).[30]

Table 1 *In vitro antiplasmodial activity (IC_{50} ± S.D.) of the extracts of selected medicinal plants against P. falciparum W2 and D6 clones, and cytotoxicity on mammalian Vero E6 cells, with Selectivity Index*

Plant species	*P. falciparum* (W2) (µg/ml)	*P. falciparum* (D6) (µg/ml)	Vero E6 cells CC_{50} (µg/ml)	Selectivity Index CC_{50}/D6
A. indica	43.70±1.45	17.90 ± 1.57	461.5± 18.7	25.8
H. abyssinica	10.25±0.30	4.40 ± 0.16	115.4 ± 4.5	25.5
M. africana	9.89±0.84	7.50 ± 1.40	252.3± 18.3	34.0
M. senegalensis	2.35±0.15	1.85 ± 0.04	235.0± 16.5	128.7
T. simplicifolia	19.89±0.84	17.40± 1.40	251.3± 18.3	117.1
W. ugandensis	3.65±0.91	2.30 ± 0.15	682.5± 18.3	299.3
Artemisinin (ng/ml)	3.38±0.49	0.90 ± 0	nd	
Chloroquine (ng/ml)	31.35±2.15	8.95 ± 1.18	nd	

The IC_{50} values are expressed as mean±S.D. of three different determinations per experiment; SI, the Selectivity Index, is defined as the ratio of the CC_{50} value determined on the mammalian cell line to that of the IC_{50} value determined on the CQ-sensitive *P. falciparum* (D6). Two reference drugs, chloroquine and artemisinin, are included as positive controls and their IC_{50} values are given in ng/ml. nd: not determined.

The root extracts of *Maytenus senegalensis* have been shown to inhibit protein kinase C, as well as having lethality against brine shrimp. The dihydroagarofuran sesquiterpene alkaloids mayteine, putterine A and putterine B have been isolated from the root extracts. Quinonemethide triterpenoids, which have revealed potential anti-tumour and anti-microbial activities, are another diverse group of secondary metabolites from the related species *Maytenus ilicifolia.*[31,32]

Various biological activities have been attributed to diverse secondary metabolites, among them maytansinoids.[33] The activity of *M. senegalensis* could presumably be ascribed to the presence of the sesquiterpenes in addition to flavonols, which collectively may be involved in the antiplasmodial activity shown in other related species.[34] It is probable that some extracts that prolonged survival time despite mild chemo-suppressive activity, or those extracts which had high chemo-suppressive activity but had less significant survival time, may have been due to mechanisms of drug action such as having an indirect effect on the immune system, or by other pathways that are not yet understood.[35]

The suppressed development of *P. berghei* in the first days of infection suggests that the extract affects the intra-erythrocytic stages of the parasite. The subsequent development of infection and shorter survival time points to a shorter duration of action of the extract, perhaps limited by rapid metabolism, distribution and excretion. The extracts of *M. senegalensis* and *W. ugandensis* had high (70.91%) and moderate (35.15%) chemo-suppression respectively. The survival time for *P. berghei*-infected mice treated with *W. ugandensis* extract was significantly longer (12.0 days) than those treated with the *H. abyssinica* extract (10.0 days).

Myrsine africana seeds are traditionally used as anthelmintics in humans and livestock by the Samburu[36] and for the treatment of digestive disorders and chest pains. The seeds are reported to contain 2,5-dihydroxyalkyl- and 2,5-dihydroxydialkylbenzoquinone derivatives, maesaquinone, acetylmaesaquinone, maesanin and embelin, of which embelin acts as an anti-feedant to *Schistocerca gregaria* and is larvicidal to *Aedes egyptii* larvae.[37,38] The methanol extract of *M. africana* exhibited moderate *in vitro* activity and high chemo-suppression (78.66%). This is the first report of *Myrsine africana* as having antimalarial activity.

The methanolic extract of *A. indica* showed moderate activity with IC_{50} =17.9μg/ml. *A. indica* has been suggested to contain active constituents which might target specific metabolically active processes at the parasitic schizont stage.[39] In a comparative study of acetone/water and aqueous extracts of the leaves of *A. indica*, they manifested inhibitory effects on a chloroquine-sensitive *P. falciparum* at a concentration of 20μg/ml.[40] Earlier findings showed that azadirachtin from *A. indica* was able to block the development of motile malaria gametes *in vitro* and raised the possibility of developing azadirachtin-based compounds as antimalarial agents with transmission blocking potential.[41] *In vitro* antimalarial activity of *A. indica* has been reported, aqueous extracts exhibiting stronger inhibitory activity than methanol extracts (IC_{50} 2.0 and 16.9 μg/ml respectively); preliminary phytochemical screening revealed terpenoids, the limonoid gedunin and flavonoids.[42]

4 CONCLUSIONS

The traditional method of preparation usually involves boiling the plant leaves or roots in water. It is rare to make a concoction of more than one part of a plant due to suspected toxicity. The extracts of *M. senegalensis* and *W. ugandensis* were highly active *in vitro*; the other water extracts were moderately active and exhibited relatively high chemo-suppression in mice infected with *P. berghei*. These were probably prodrugs, metabolized into active constituents *in vivo*. This may confirm the consistent claims by traditional healers that these plants are effective in treating malaria in humans. Their antiplasmodial activities and relative safety further confirm their ethnophytotherapeutic usefulness as antimalarials.

The plant extracts with IC_{50} <10μg/ml offer the potential for the isolation of lead antimalarial compounds, or for the characterization of some active compounds that could be used as markers for standardization of the extracts for use as traditional antimalarials. We suggest that some Samburu plants used in traditional medicine possess potent antiplasmodial activity with minor effects on mammalian cells and hence we propose that those plants should be further investigated to contribute to the development of potential antimalarial medicines. The results of the preliminary studies carried out with the above-mentioned plants are encouraging.

The scope for developing antimalarials from indigenous plants depends on the screening of a large number of medicinal plants from different geographical regions, especially from the rural areas, where usage of these medicinal plants is more common. Once the antiplasmodial effect of the plant is confirmed, the active ingredients could be isolated by different extraction methods.

Acknowledgements

We thank Mr. Adamson Lanyansuya of SIRAN for facilitation of the field studies, the Samburu people for their cooperation and Mr. Musembi Kimeu of LARMAT herbarium, University of Nairobi, for assisting in the botanical identification.

References

1 WHO (World Health Organization) Centre for Health Development, 2002.
2 D.A. Fidock, P.J. Rosenthal, S.L. Croft, R .Brun and S. Nwaka, *Nature Rev. Drug Discov.*, 2004, **3**, 509.
3 P.L. Olliaro and P.B. Bloland, *Drug Discovery*, Humana Press, Totowa, NJ, 2001, p. 65.
4 C.N. Muthaura, G.M. Rukunga, S.C. Chhabra, S.A. Omar, A.N. Guantai, J.W. Gathirwa, F.M. Tolo, P.G. Mwitari, L.K. Keter, P.G. Kirira, C.W. Kimani, G.M. Mungai and E.N. Njagi, *J. Ethnopharmacol.*, 2007, **112,** 545.
5 S. Bah, A.K. Jäger, A. Adsersen, D. Diallo and B.S. Paulsen, *J. Ethnopharmacol.*, 2007, **110**, 451.
6 J.W. Gathirwa, G.M. Rukunga, E.N. Njagi, S.A. Omar, P.G. Mwitari, A.N. Guantai, F.M Tolo, C.W. Kimani, C.N. Muthaura and P.G. Kirira, *J. Ethnopharmacol.*, 2008, **17**, 223.
7 J.F. Trape, *Trends Parasitol.,* 2002, **18**, 224.
8 J. Bickii, N. Njifulie, J.A. Foyere, L.K. Basco and P. Ringwald, *J. Ethnopharmacol.*, 2000, **69**, 27.
9 C. Geoffrey and G.C. Kirby, *Trans. R. Soc. Trop. Med. Hyg.,* 1996, **90**, 605.
10 M.O. Nanyingi, J.M. Mbaria, A.L. Lanyasunya, C.G. Wagate, B.K. Kipsengeret, H.F. Kaburia, R.W. Munenge and W.O. Ogara, *J. Ethnobiol. Ethnomed.,* 2008, **4**,14.
11 O. Kayser, A.F. Kiderlen and S.L. Croft, *Parasitol. Res.*, 2003, **90**, 55.
12 Q. Le Tran, Y. Tezuka, J. Ueda, N.T. Nguyen, Y. Maruyama, K. Begum, H.S. Kim, Y. Wataya, Q.K. Tran and S. Kadota, *J. Ethnopharmacol.*, 2003, **86**, 249.
13 G.N. Wanyoike, S.C. Chhabra, C.C. Lang'at-Thoruwa and S.A. Omar, *J. Ethnopharmacol.,* 2004, **90**, 129.
14 F.W. Muregi, A. Ishih, T. Miyase, T. Suzuki, H. Kino, T. Amanod, M. Gerald, G.M. Mkoji and M. Terada, *J. Ethnophamacol.*, 2007, **111**, 190.
15 M.G.L. Brandao, T.S.M. Grandi, E.M.M. Rocha, D.R. Sawyer and A.U. Krettli, *J. Ethnopharmacol.*, 1992, **36**, 175.
16 J.O. Kokwaro, in *Medicinal Plants of East Africa*. East African Literature Bureau, Nairobi, 1977, p 380.
17 H.J. Beentje, in *Kenya Trees, Shrubs and Lianas*, National Museums of Kenya, Nairobi, Kenya. 1994, p 722.
18 M. Kurokawa, T. Hozumi, M. Tsurita, S. Kadota, T. Namba and K. Shiraki, *J. Pharmacol. Exp. Ther.,* 2000, **297**, 372.
19 WHO/EDM/TRM/2000.1, in *General Guidelines for Methodologies on Research and Evaluation of Traditional Medicines.* WHO, Geneva, 2000, p 74.
20 W. Trager and J.B. Jensen, *Science*, 1976, **193**, 673.
21 A.M. Kaou, V. Mahiou-Leddet, S. Hutter, S. Sidi Aïnouddine, S. Hassani, I.Yahaya, N. Azas and E. Ollivier, *J. Ethnopharmacol.*, 2008, **116**, 74.
22 N. Azas, N. Laurecin, F. Delmas, C. Di Giorgio, M. Gasquet, M. Laget and P. Timon-David, *Parasitol. Research*, 2002, **88**, 165.
23 D.G. Sixsmith, W.M. Watkins, J.D. Chulay and H.C. Spencer, *Am. J. Trop. Med .Hyg.*, 1984, **33**, 772.
24 W. Peters, J.H. Portus and B.L. Robinson, *Ann. Trop. Med. Parasitol.*, 1975, **69**, 155.
25 L. Tona, K. Mesia, N.P. Ngimbi, B. Chrimwami, Okond'ahoka, K. Cimanga, T. De Bruyne, S. Apers, N. Hermans, J. Totte, L. Peters and A.J. Vlietinck, *Ann. Trop. Med. Parasitol.*, 2001, **95**, 47.
26 H.A.A. Twaij, A. Kery and N.K. Al Khazraji, *J. Ethnopharmacol.*, 1983, **9**, 299.

27 P.O. Obih and J.M. Makinde, *Afr. J. Med. Sci.*, 1985, **14,** 55.
28 M.C. Gessler, M.H.H. Nkunya, J. Chollet, M. Heinrich and M. Tanner, *Phytother. Res.*, 1995, **9,** 504.
29 A. El-Tahir, G.M.H. Satti and S.A. Khaild, *J. Ethnopharmacol.*, 1999, **64**,.227.
30 C. Clarkson, J.M. Vinesh, R.C. Neil, M.G. Olwen, P. Pamisha, G.M. Motlalepula, B. Niresh, J.S. Peter and I.F. Peter, *J. Ethnopharmacol.*, 2004, **92**, 177.
31 B.T. Schaneberg, K.D. Green and A.T. Sneden, *J. Nat. Prod.*, 2001, **64**, 624.
32 W. Buffa Filho, J. Corsino, S.V. da Bolzani, M. Furlan, A.M. Pereira and S.C. Franca, *Phytochem. Anal.*, 2002, **13**, 75.
33 P.J. Reider and D.M. Roland, in *The Alkaloids*, Academic Press, New York, 1984, p.71.
34 L.O.A. Manguro, I. Ugi, R. Hermann and P. Lemmen, *Phytochemistry,* 2003, **63**, 497.
35 P. Rasoanaivo, A. Petitjean, S. Ratsimamanga-Urverg and A. Rakato-Ratsimamanga, *J. Ethnopharmacol.*, 1992, **37**, 117.
36 J.M. Gathuma, J.M. Mbaria, J. Wanyama, H.F.A. Kaburia, L. Mpoke, J.N. Mwangi and Samburu and Turkana healers, *J. Ethnopharmacol.*, 2004, **91**, 7.
37 J.O. Midiwo and L.M. Arot, *Natural Prod. Lett.*, 1996, **8**, 11.
38 J.O. Midiwo, R.W. Mwangi and Y. Ghebremeskel, *Insect Sci. Appl.,* 1995, **16**, 163.
39 R. Dhar, K. Zhang, G.P. Talwar, S. Grag and N. Kumar, *J. Ethnopharmacol.*, 1998, **61**, 31.
40 J. Iroka, *Trans. R. Soc. Trop. Med. Hyg.*, 1993, **87**, 471.
41 I.W. Jones, A.A. Denholm, S.V. Ley, H. Lovell, A. Wood and R.E. Sinden, *FEMS Microbiol. Lett.,* 1994, **120**, 267.
42 I. Muhammad, K.A. El Sayed, J.S. Mossa, M.S. Al-Said, F.S. El-Feraly, A.M. Clark, C.D. Hufford, S. Oh and A.M. Mayer, *J. Nat. Prod.*, 2000, **63**, 605.

ANTI-LARVAL ACTIVITY OF CRUDE PLANT EXTRACTS FROM *VITEX SCHILIEBENII* AND *VITEX PAYOS* (VERBENACEAE)

G N. Mokua,[1] I.O. Ndiege,[1,2] A. Hassanali[2] and P.K. Tarus[1]

[1]Chemistry Department, Kenyatta University, P.O. Box 43844, Nairobi, Kenya.
[2]Behavioral and Chemical Ecology Department, I.C.I.P.E, P.O. Box 30722, Nairobi, Kenya.

1 INTRODUCTION

Verbenaceae is a tropical and sub-tropical family of about 75 genera and 3000 species. It includes herbs, shrubs, lianas and trees.[1] Many species of this family are used for a number of medicinal, timber and food purposes. For instance, timber from *Citharexylum* species is used to make musical instruments, while *Vitex agnus castus* is renowned for its medicinal properties.[2]

Shrubs of the genus *Vitex* in the family Verbenaceae are found mainly in tropical and sub-tropical regions, although a few species may be found in temperate zones.[3] The origin of the name of the genus is obscure,[4] though it may be from a Latin word meaning "to blind",[5] referring to the flexible twigs.[6]

The local members of this genus include *Vitex doniana* (black plum, mfudu in Kiswahili, mufuku in Gikuyu and mutahuru in Ekegusii); *Vitex mombassae* (mfudumaji in Kiswahili, mfududu in Giriama and mfudukoma in Digo); *Vitex payos* (chocolate berry, mfufu in Kiswahili and mfudu in Digo); *Vitex amboniensis* (mufudu in Giriama); *Vitex fischeri* (mohutu in Nandi and mufutumwe in Ateso); *Vitex keniensis* (Meru Oak, muhuru in Gikuyu and muuru moru in Kimeru); *Vitex schiliebenii; Vitex madiensis*; *Vitex ferruginea*; *Vitex strickeri* (mukichano in Giriama, mvumbain in Kiswahili and Mukakinga in Gikuyu), *Vitex zanzibariensis* (vifuu in Kiswahili) and *Vitex tangensis* (mgegi in Kiswahili and mfududu/mufudumaji in Giriama).

Vitex payos (chocolate berry) is a deciduous tree that may grow up to 8 m high with a low rounded crown. Its bark is grey-brown and deeply fissured. Younger stems have a rusty, woolly bark and the leaves have five leaflets. In Kenya, it grows in Kitui, Embu, Machakos, Kilifi and Kwale in low, hot and semi-arid lands at 0-1600 m above sea level.[7] A decoction of the root has been reported as a remedy for stomach problems, the pounded bark is administered to treat threadworm and skin problems, while boiled leaves are taken by patients to improve their appetite for food.[8] *Vitex schiliebenii* is a branched shrub with multiple stems and low height (4-8 m). The leaves have three leaflets with a smooth surface, about 10-12 cm long. In Kenya, it grows in the coastal region at Watamu, which is a low and semi-arid land.

Plants of the genus *Vitex* exhibit insecticidal activities. *Vitex negundo* extracts exhibit larvicidal activity against *Culex quinquefasciatus* and *Anopheles stephensis*,[9] and act as a feeding deterrent to *Aedes aegypti* larvae.[10] *Vitex rotundifolia* also shows feeding deterrent properties towards *Aedes aegypti* larvae, while insecticidal substances were reported to be present in the leaf tissues.[11] Rotundial, isolated from *Vitex rotundifolia*, exhibited insect repellent properties. *Vitex trifolia* has insecticidal properties and components of the seeds have anti-feedant activity.[12]

In this paper we report the larvicidal activity of the crude extracts of the stem barks of *Vitex schiliebenii and Vitex payos*.

2 METHOD AND RESULTS

2.1 Collection and extraction of plants

The stem barks of *Vitex schiliebenii* and *Vitex payos* were collected from Watamu (Malindi District) and North Coast (Kilifi District), respectively. The plant materials were dried under shade for 7 days and ground into powder. The dry powder (2 kg) was soaked in 5 litres of the solvents indicated below for 3 days, with occasional stirring. The solution was decanted and the process was repeated 3 times. The solvent extracts were combined, filtered and concentrated at low temperature and pressure in a rotary evaporator. The extraction was done sequentially in solvents of increasing polarity (*n*-hexane, chloroform, ethyl acetate and methanol) to obtain different extracts on which to perform the bioassay.

2.2 Larvicidal assay

A stock solution was prepared by dissolving 1.25 mg of each extract in 1 ml of solvent (5 % DMSO in acetone for the hexane, chloroform and ethyl acetate extracts, and water for the methanol extract) to obtain concentrations of 1250 ppm. The stock solution was diluted to 25, 12.5, 6.25, 2.5 and 1.25 ppm. Freshly moulted late 3rd-instar *Anopheles gambiae* larvae (20) were released into each test beaker containing 100 ml of distilled water treated with the respective sample concentration. In the control experiments, the larvae were exposed to 2.0, 1.0, 0.5, 0.2 and 0.1 ml acetone diluted to 100 ml with distilled water. Fish food was provided for the larvae, and the test beakers were placed in a room maintained at 26°C and with a photoperiod of 12 hours light and 12 hours darkness. Mortality of the larvae was monitored after every 24 hours until the death of the last larva or pupa or their emergence into adults. The experiment was replicated 5 times for each concentration. Mortality was calculated according to the formula % mortality = 100Y/Z, where Z = initial number of larvae introduced into each test beaker and Y = mean death defined by the difference between the mean test deaths and the mean control deaths.

Table 1 shows the lethal concentration (LC) values for the extracts of *Vitex schiliebenii* after 72 hours. The methanol and n-hexane extracts show interesting LC_{50} values, of which the methanol extract was the more active, with LC_{50} of 4.9 ppm.

Table 1 *Larvicidal activity data for* Vitex schiliebenii *extracts*
Lethal concentrations (LC) (ppm) after 72 h

Extract	LC_{50}	LC_{75}	LC_{90}
C_6H_{14}	5.5	11.7	17.2
$CHCl_3$	10.4	18.0	24.9
EtOAc	9.9	16.1	21.7
MeOH	4.9	8.6	12.0

Table 2 shows the lethal concentration (LC) values for the extracts of *Vitex payos* after 72 hours. The n-hexane extracts showed the highest activity (LC_{50} 3.1 ppm), while the chloroform extracts also showed good activity (LC_{50} 8.9 ppm).

Table 2 *Larvicidal activity data for* Vitex payos *extracts*
Lethal concentrations (LC) (ppm) after 72 h

Extract	LC_{50}	LC_{75}	LC_{90}
C_6H_{14}	3.1	11.6	19.3
$CHCl_3$	8.9	12.2	15.2
EtOAc	14.8	26.2	36.4
MeOH	9.7	27.1	42.7

2.3 Discussion

All the larvae were normal and active immediately after exposure to each test compound at all the concentrations tested, and the normal feeding process and zigzag motions of the larvae could be seen (Plate 1). However, after 24 h exposure to 1.25-12.5 ppm of the extracts, some abnormal sluggishness was observed in the test experiments with the larvae sinking down and floating again. After 48-192 h, some larvae were excited or restless, while others died (Plate 2). Morphological disruption of the extension on the terminal segment was observed in the treated and dead larvae (Plates 3 and 2, respectively). The treated larvae also showed signals of a swollen thorax and abdomen.

Although all the extracts from the two plants showed interesting larvicidal activities, it was the methanol extract of *Vitex schiliebenii* that exhibited the highest activity. The *n*-hexane extracts of the same plant also had good activity. These biological effects can probably be attributed to the presence of certain secondary metabolites in the extracts. Consistent with this suggestion is the observation that mortality of the larvae of *Anopheles gambiae* exposed to the plant extracts increased with the concentration of the extracts (1.25-25 ppm) and the time for which the insects were exposed to them. Similar observations have previously been reported for larvae of *Culex quinquefasciatus* and other mosquito species exposed to the extracts of several other plants such as *Nerium indicum* and *Euphorbia royleana*.[14]

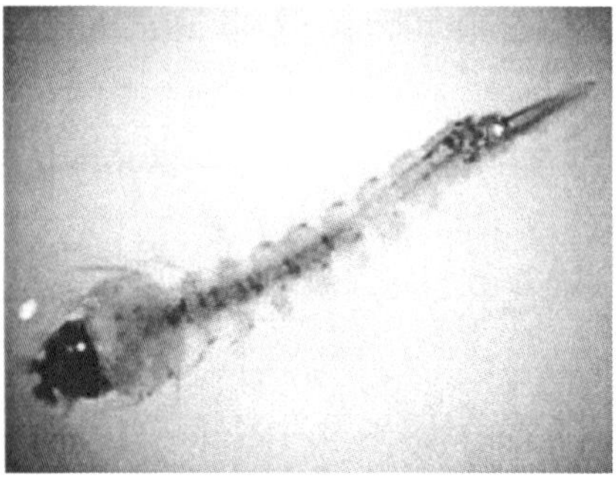

Plate 1 *Untreated larva (control)*

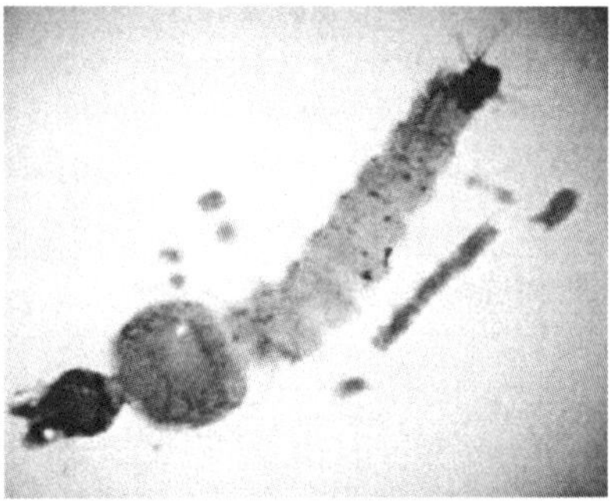

Plate 2 *Treated dead larva*

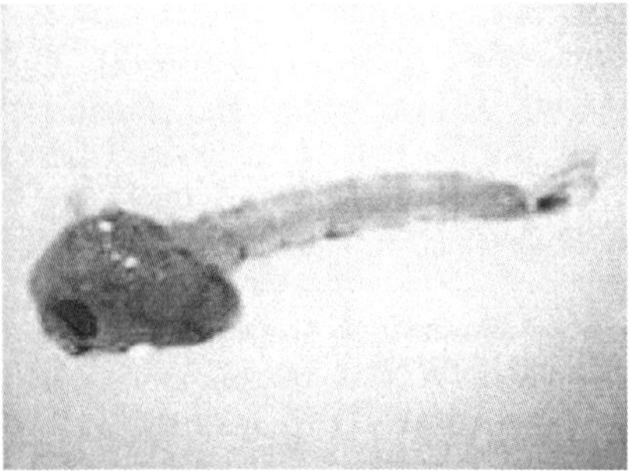

Plate 3 *Larval-pupal intermediate*

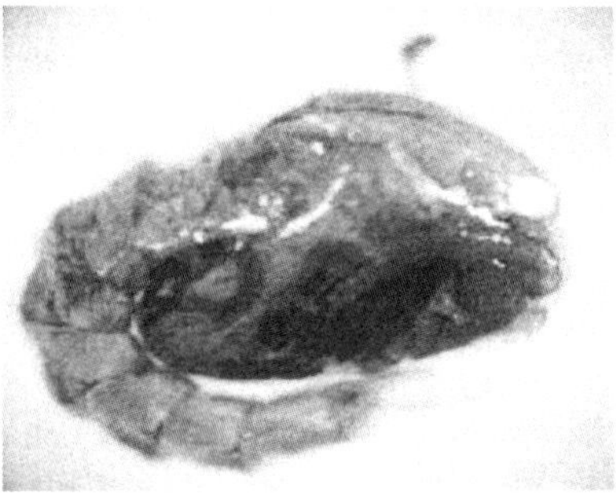

Plate 4 *Pupa from larvae emerging to adults*

The extracts also showed useful effects at lower concentrations. It was observed that 20-40% of the larvae treated with the low concentration of the extract successfully pupated but often died before emerging into adults. This is no disadvantage for malaria control programmes since it is the adult *Anopheles gambiae* mosquitoes that are the prime vectors of the disease.

As well as the direct toxicity of the extracts, morphological abnormalities were observed in 15-20% of the larvae exposed to the plant extracts. Under the light microscope, both treated and control larvae showed similarities in physical morphological architecture. However, distinct differences were observed in the structure of the extension on the terminal segment, which was darker in the treated than in the normal (control) larvae. The abdomens of the treated larvae were also swollen, and larval-pupal intermediates, with pupal heads and larval abdomens, were also observed amongst the treated larvae. The structural deformation of the abdomen and consequently of the anal gills probably led to their dysfunction, and this may have led to the death of the mosquito larvae.[15] In fresh-water, mosquito larvae take up and eliminate most ions via the anal gills, while the organs responsible for retaining a balance of essential ions is mainly located in the alimentary canal.[15] The dysfunction of the anal gills probably led to an interruption of the osmotic and ion regulation which resulted in the swelling of the abdomen.[15]

Acknowledgements

Financial support from WHO/TDR/MIM Project (A 10628) through ICIPE is gratefully acknowledged. We also thank ICIPE insectary staff for providing the mosquito larvae for the bioassay.

References

1 A.L. Poole and M.A. Nancy, *Trees and Shrubs of New Zealand*. Manaaki Whenua Press, New Zealand, 1994.
2 J.T. Salmon, *Native New Zealand Flowering plants*. Reed, Auckland, New Zealand, 1991.
3 X. McMillan, *A Concise Dictionary of Plants Cultivated in the United States and Canada*, L.H. Bayley (ed.)., Cornell University Press, New York, 1976, p. 1161.
4 A. Eagles, *Trees and Shrubs of New Zealand*, Vol.1, Revised Edn., Collins, Auckland, New Zealand, 1986.
5 H. Bruce, *Nature in New Zealand. Native Trees*. Collins, Auckland, New Zealand, 1985, p 70.
6 J.W. Mathews, *New Zealand Trees, a Popular Guide to the Identification of the Principal Species*. A.H. Wellington and A.W. Reed, Auckland, New Zealand, 1962.
7 P.N. Maundu, W.G. Ngugi and H.S. Kabuye, *Traditional Food Plants of Kenya*. Kenya Resource Centre for Indigenous Knowledge, National Museums of Kenya, Nairobi, 1999, pp. 241-244.
8 K. Christopher, A.B. Ruffo and T. Bo, *Edible Wild Plants of Tanzania*. Regional Lands Management Unit (RELMA), SIDA, Nairobi, Kenya, 2002.
9 E. Pulshpalatha and J. Muthukrishnan, *Indian J. Malariol.*, 1995, **32**, 14.
10 D.S. Hebbalkar, G.D. Hebbalkar, R.N. Sharma, V.S Joshi and V.S. Bhat, *Indian J. Med. Res.*, 1992, **95**, 200.
11 K. Watanabe, Y. Takada, N. Matsuo and H. Nishimura, *Biosci. Biotechnol. Biochem.*, 1995, **59**, 1979.
12 S. Hosozawa, N. Kato, K. Muakata and Y.L. Chen, *Agric. Biol. Chem.*, 1974, **38**, 1045.
13 W.S. Abbott, *J. Econ. Entomol.*, 1925, **18**, 265.
14 W. Choochote, B. Tueton, D. Kanjanapothi, E. Rattanachanpichoi, U. Chaithong, A. Jitpakdi, P. Tippawangkosol, D. Rivong and B. Pittasawat, *J. Vector Ecol.*, 2004, **29**, 340.
15 A.N. Clements, *The Biology of Mosquitoes: Development, Nutrition and Reproduction*. Vol. **1**. Chapman & Hall, London, 1992, pp. 54-60.

NATURAL PRODUCTS AS LEADS FOR NEW AND INNOVATIVE CROP PROTECTION CHEMICALS

J.M. Clough

Syngenta, Research Chemistry, Jealott's Hill International Research Centre, Bracknell, Berkshire, RG42 6EY, UK

1 INTRODUCTION

This paper describes the way in which Syngenta, a global crop protection company, uses biologically-active natural products as leads for the invention of new herbicides, fungicides and insecticides.

The world market for crop protection chemicals is large, with sales worth $33.4 billion in 2007. Herbicides are the most valuable sector, worth 48.3% of the market in 2007, followed by fungicides and insecticides, valued at 24.3% and 24.0%, respectively. The remaining 3.4% of the market comprises products such as plant growth regulators. The market is mostly in Europe, North America, Asia and, increasingly, in Latin America, while the market in Africa is still relatively small.[1]

Each new crop protection chemical costs in the order of $200 million to invent, develop, register and launch. This high cost is justified only for products of substantial value, and it is not easy to find new compounds of this sort, with all the properties required for a modern crop protection chemical. Such a compound must have high but selective potency, in order to control a useful spectrum of weeds, insects or fungi when applied at low rates, and without affecting non-target organisms. It must have sufficient chemical and metabolic stability to survive in sunlight on a leaf surface and then within the target organism, but not be so stable that it persists in the environment. It requires physical properties which allow it to reach the enzyme or receptor at which it acts, and then to bind efficiently. And finally it must be cost-effective, which usually means that it must be relatively cheap to manufacture, although chemicals with extremely high potency can support higher manufacturing costs.

2 THE INVENTION OF CROP PROTECTION CHEMICALS FROM NATURAL PRODUCT LEADS

The first stage in the discovery of a new crop protection chemical is lead generation, in which the chemist needs to find an active compound on which to work. This compound is likely to be only weakly active, and it may have other defects, such as poor selectivity, for example, but it represents a starting point for a programme of analogue synthesis. One option is to analyse the recent patent literature, looking at the chemistry being carried out by competitor companies in the hope that loopholes can be found. Another is to conduct further synthesis in areas of chemistry and around modes of action that have been fruitful in the past, in the expectation that new and useful products will be found. Still another is to screen sets of compounds, obtained by synthesis, or from external suppliers, against plants, fungi and insects, looking for actives. Finally, and this is the subject of this paper, the chemist can look to natural products for inspiration. Although it is difficult to turn natural products into useful commercial products, they do represent a source of real novelty, in terms of both structure and mode of action.

Natural products lie at the very origins of organic chemistry. Early chemists were intrigued by the pigments, toxins and essential oils that they could isolate from natural sources such as plants and fungi. Little by little, they developed methods to characterise and then synthesise these chemicals. Natural products still play a pivotal role in organic chemistry. New examples continue to be isolated, and the need to characterise and synthesise novel structural types still drives the development of new analytical and synthetic methods. The question for those of us working in the crop protection industry is how to make the best use of these compounds in the invention of new herbicides, fungicides and insecticides.

Natural products can occasionally be used for crop protection purposes in their own right. The natural pyrethrins, such as pyrethrin I (Figure 1), are insecticidal compounds produced in the flowers of the plant *Chrysanthemum cinerariaefolium*, and are well-known examples. They are used to control insect pests on various crops and on farm and domestic animals, but their use outdoors is limited by their poor stability in sunlight.[2,3] Incidentally, this example is highly appropriate for this conference held at the University of Nairobi, because much of the world's supply of pyrethrins is produced in Kenya.

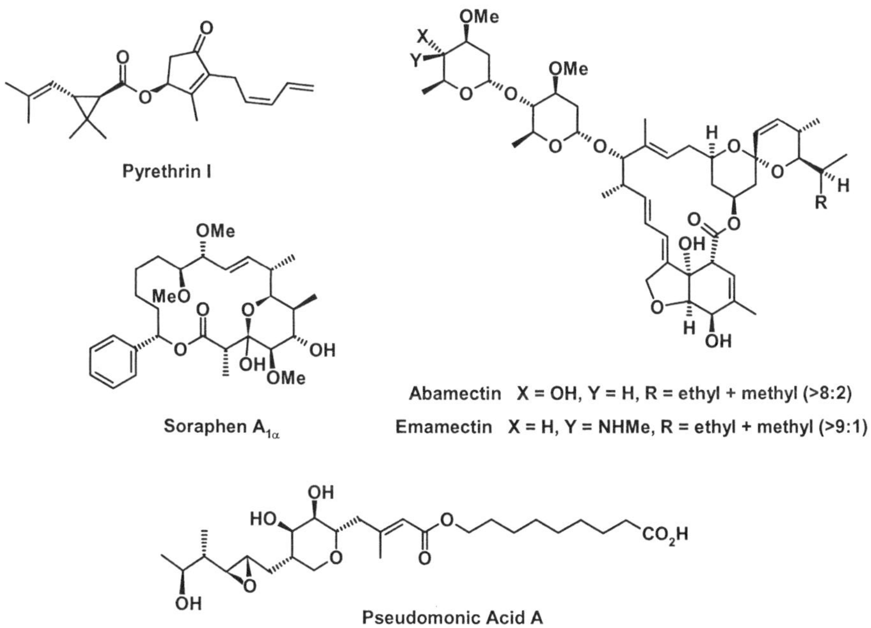

Figure 1 *Examples of natural products that have inspired crop protection research*

Natural products can sometimes be used in crop protection following suitable chemical modification. A good example of a semi-synthetic compound is Syngenta's insecticide emamectin, prepared by synthetic modification of the natural product abamectin (= avermectin B_1) which, in turn, is obtained by fermentation of the bacterium *Streptomyces avermitilis* (Figure 1). Emamectin is a highly potent insecticide, used at rates as low as 5-25 g/ha, and it is this that justifies the high costs associated with its production.[4]

Another use of natural products, sometimes overlooked, is their value in validating novel modes of action. For example, it was discovered that certain close analogues of pseudomonic acid A have herbicidal activity, and that this activity is the result of inhibition of isoleucyl-tRNA synthetase (ITRS) in plants.[5] This was the first time that inhibitors of this enzyme had been shown to have herbicidal activity, a result that validated the enzyme as a herbicide target, indicating that other inhibitors of the enzyme, assuming they had suitable physical properties, would also be expected to have herbicidal activity. Similarly, the fungicidal natural product soraphen $A_{1\alpha}$ validated acetyl-CoA carboxylase as a fungicide target.[6]

But most important for Syngenta is the use of natural products as the inspiration for programmes of synthesis, and our strategy is to focus on this approach, which leads ultimately to fully synthetic commercial products. Over the years, we have worked on many projects based on natural products. Most, of course, have not been successful, but a small number have led to successful commercial products. Examples are described in the second half of this paper.

Syngenta is always looking for new bioactive natural products with attractive structures as starting points for synthesis, and we obtain these in several ways. One is to carefully follow the scientific literature, looking for reports of new natural products with suitable biological activity and then, whenever possible, obtaining samples for testing in our own assays. It is also sometimes possible to purchase samples of pure natural products, and these usually have to be selected solely on the basis of their structures, since bioactivity data are rarely available. Most importantly of all, we have a major natural products discovery programme with partners in China. This is a true partnership because each party has skills that it shares with the other, and each will share in the profits from the products discovered.

We have found that a properly designed and well-run bioassay-directed discovery programme is still a good way to find new natural products. We have chosen to use micro-organisms rather than plants because there is an almost limitless supply of different strains, and scale-up and re-supply is more straightforward. Micro-organisms are allowed to ferment and extracts of the resulting broths are tested for activity against a variety of plants, fungi and insects.

Of course, there are intrinsic disadvantages with this approach. For example, the re-discovery of known compounds is common, so it is important to have a suitable "de-replication" step early in the process. Active extracts are therefore compared by HPLC-MS with our database of natural products, and those containing compounds already known to us are discarded. Another disadvantage is that substantial time and effort are invested before the natural product is finally isolated and its structure and bioactivity are determined, and the result can often be disappointing. However, the expectation is that the discovery of a small proportion of exciting natural products will make the whole exercise worthwhile.

One feature of a natural product discovery programme is that the compounds emerging often have rather weak activity, and it is difficult to pick out those that have the potential to be turned into commercial products. Only by synthesising and testing carefully designed analogues can we tell which might be developed into high quality lead projects.

Synthesis resource is always stretched (it is the bottleneck in many projects) and natural products must compete with leads which are often considered to be more reliable starting points for synthesis. As a result, only a carefully selected subset of bioactive natural products can be worked on. How are these chosen?

First of all, natural products with functional groups associated with chemical reactivity or toxicity can be discarded. For example, many are phenolic, and these are generally unpromising because they are prone to oxidation and conjugation within a plant. Quinones and other redox-reactive compounds are also unattractive, as are natural products which are likely to be metal-chelators, such as hydroxamic acids and catechols.

Next, the natural product should have an appropriate degree of structural complexity. Those that are highly complex are difficult targets for synthesis (although the strategy might be to prepare simplified analogues and part structures in the hope that these will retain the activity of the lead compound). On the other hand, those that are too simple are also unattractive because it is difficult to see which analogues to make, and because such compounds may not be binding at a specific enzyme site, in which case experience tells that it will be difficult to turn them into potent and selective products.

A collection of enzymes representing most of the known herbicide, fungicide and insecticide modes of action is also available to us. By testing a new natural product against these enzymes, it can be determined whether it has a known or novel mode of action. Those with a novel mode of action are particularly attractive since they may give new biological effects and should be resistance-breaking. It is also useful to know something about the chemical and metabolic stability of a new natural product, e.g. its stability to hydrolysis, oxidation, light, or the nucleophile glutathione, and on exposure to plant cell cultures. If the compound has biological activity despite being rather unstable, this adds to its interest since analogues that are designed to be more stable should have higher potency.

So, armed with such information, the chemist defines his programme of synthesis. He needs to address certain questions, the most fundamental of which is whether the structure of the natural product can be simplified and its activity (potency and spectrum) improved. To put it another way, which structural features of the natural product are required, and which can be removed without loss of activity? Many projects of this type have been conducted within Syngenta and its parent companies over the years, and several of these are briefly described in the following section.

3 CASE STUDIES

3.1 The herbicidal pseudomonic acids[5,7]

The pseudomonic acids are a family of naturally-occurring antibiotics produced by fermentation of a strain of the bacterium *Pseudomonas fluorescens*. Pseudomonic acid A, the major metabolite, has a broad spectrum of activity against Gram-positive bacteria and is marketed for topical use in the treatment of skin infections (Figure 2). It acts through the inhibition of protein biosynthesis by selectively binding to the bacterial isoleucyl-tRNA synthetase (ITRS). Inhibition of the corresponding human enzyme is four orders of magnitude lower.

Pseudomonic acid A can be hydrolysed to monic acid A, which can then be converted into a variety of semi-synthetic esters and amides, many of which are herbicidal because they inhibit ITRS in plants. As noted above, this was the first time that herbicidal inhibitors of this enzyme had been discovered, so this important observation validated a new mode of action for herbicides. Even the simple methyl ester of monic acid A has some activity, but the 2-(vinyloxy)ethyl ester (1) is much more active, and was taken forward into extensive field trials in cereals. In the best of these, it gave good control of several important weeds at rates of 25-50 g/ha, but its performance varied depending on the growth stage and species of weed, as well as on environmental conditions, and it was not commercialised. The trifluoroethyl amide (2) is an example of a highly active amide derivative.

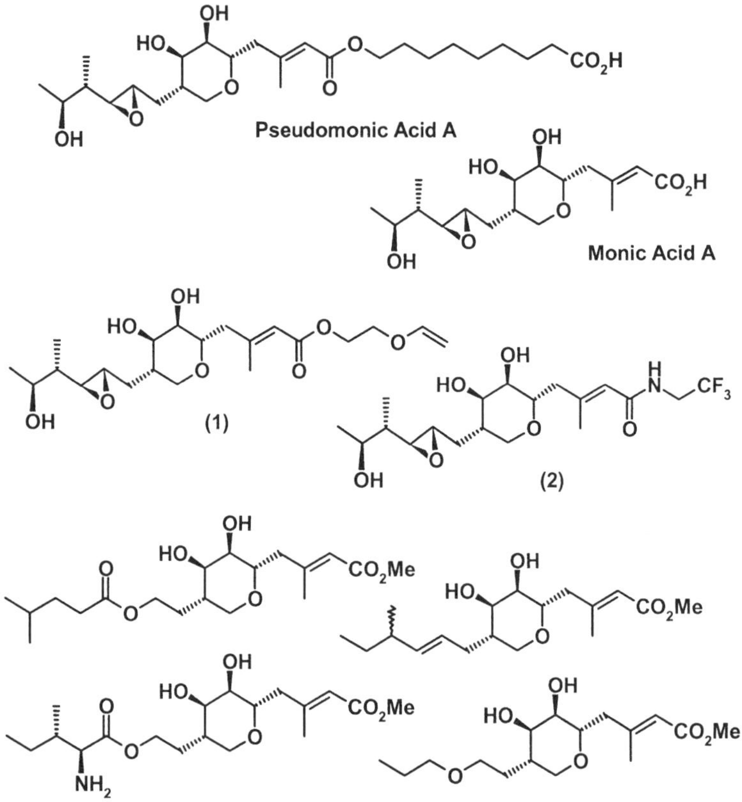

Figure 2 *Pseudomonic acid A and synthetic analogues*

A variety of fully synthetic analogues of pseudomonic A were also prepared. The pharmaceutical industry had already published a series of patent applications claiming antibacterial analogues of the natural product in which the "right-hand" (ester) side-chain had been modified, and we found that some of these also had herbicidal activity. But we focused our attention on the synthesis of analogues in which the left-hand (epoxide) side-chain of the natural product had been modified. This side-chain contains four of the eight chiral centres of the natural product, so there appeared to be good scope for simplification.

The use of divergent intermediates allowed the rapid synthesis of a diverse collection of compounds, four examples of which are shown at the bottom of Figure 2. Although the best showed some inhibition of ITRS from plants, none had significant activity in the glasshouse. The left-hand side-chain of pseudomonic acid A bears some resemblance to isoleucine, and it has been speculated that it binds at the isoleucine recognition site of the enzyme. Clearly, this site needs to be very tightly defined, since it is responsible for distinguishing between isoleucine and the other closely-related proteinogenic amino acids.

3.2 The insecticidal stemofolines[8]

Stemofoline is an insecticidal alkaloid produced in the leaves and stems of the plant *Stemona japonica*. It has a good spectrum of activity, with rapid action, and is a potent agonist of the nicotinic acetylcholine receptor in insects. These features make it a promising lead, except that its highly complex polycyclic structure is a real challenge for synthesis (Figure 3).

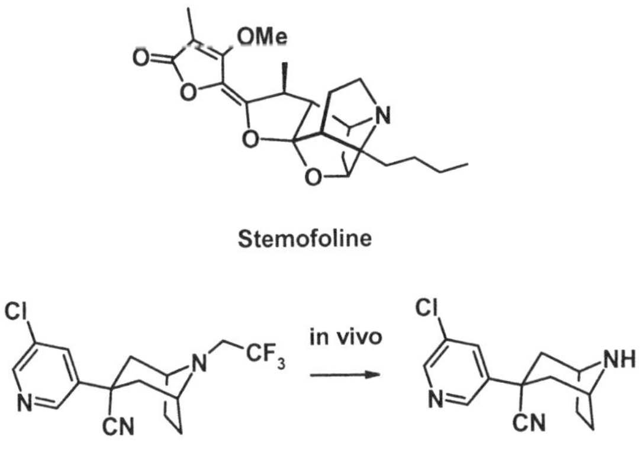

(3) Procide

pKa ~ 8.5

Figure 3 *Stemofoline and a synthetic analogue*

A breakthrough came with the recognition of a tropane substructure embedded within the framework of stemofoline. During the exploration of various tropanes, a series of highly active cyanotropanes was discovered. A leading example is the compound (3), which has a good spectrum of insecticidal activity, with high potency against aphids and whitefly and rapid action both by contact and stomach routes. Furthermore, the compound shows good xylem systemicity in plants. Compound (3) is a procide. The trifluoroethyl group is cleaved off within the insect to unmask a secondary amine which is the true active moiety. The importance of the trifluoroethyl group is that it reduces the basicity of the active secondary amine, which has pKa ~ 8.5 and would therefore be largely protonated at physiological pH to give a species with poor transport properties in plants. None of the cyanotropanes has been commercialised.

3.3 The strobilurin fungicides[9-11]

Strobilurin A and oudemansin A are the simplest members of a family of fungicidal natural products produced by several species of Basidiomycete fungi, e.g. the porcelain fungus, *Oudemansiella mucida*, that grow on rotting wood (Figure 4). We found these natural products attractive because of their simple structures, and because they have a novel mode of action, inhibiting mitochondrial respiration by binding at a site on cytochrome b, a component of Complex III. However, strobilurin A breaks down rapidly in light, and this was clearly a factor that needed to be addressed in any programme of synthesis.

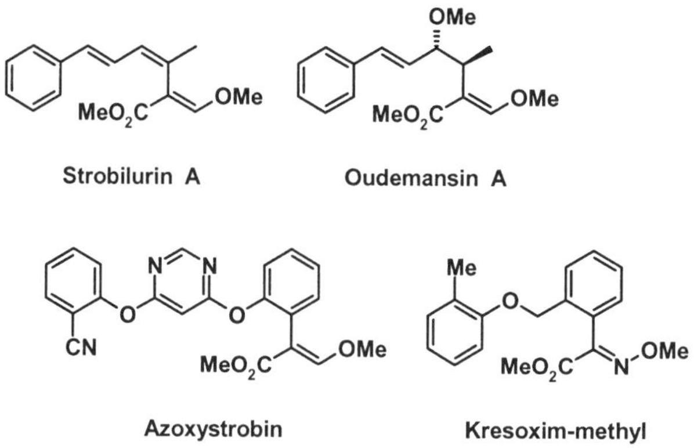

Figure 4 *The commercial synthetic agricultural fungicides azoxystrobin and kresoxim-methyl, inspired by the natural strobilurins, such as strobilurin A and oudemansin A*

Independent programmes of synthesis were carried out by one of Syngenta's parent companies and by BASF, leading to the discovery of azoxystrobin and kresoxim-methyl, respectively. On publication of the first patent applications, other companies realised the importance of the strobilurins and began their own research programmes in the area. Over a period of several years, this has led to other commercial strobilurins and there is now a family of products from different companies, and the strobilurins are one of the most important classes of agricultural fungicide. Azoxystrobin is the biggest-selling strobilurin and, indeed, the world's most valuable agricultural fungicide, with sales well in excess of $500 million per annum.

3.4 The fungicidal crocacins[12]

Crocacin D (Figure 5) has the highest fungicidal activity of the four crocacins produced by the myxobacterium *Chondromyces crocatus*. It acts through the inhibition of mitochondrial respiration in fungi by binding at a site on Complex III. Importantly, the active site is different from that at which the strobilurins bind, and crocacin D is not cross-resistant with the strobilurins.

Although the novel structure and mode of action of crocacin D were attractive, it was found to break down quickly on exposure to an esterase (hydrolysis of the methyl ester) or to light. A programme of synthesis was therefore initiated, to search for more stable analogues with simplified structures and higher potency. Examples of the compounds prepared are shown in Figure 5. Certain compounds such as (4) in which the complex "left-hand" side-chain had been replaced with a diphenyl ether group retained activity both in the *in vitro* respiration assay and against vine downy mildew growing on vines in the glasshouse. On the other hand, no active compounds in which the (*Z*)-enamide group had been replaced with an *ortho*-disubstituted benzene ring, such as (5), could be found. And, furthermore, compounds such as (6) in which the glycine ester group had been replaced in order to improve hydrolytic stability were, at best, only weakly active. No compounds of potential commercial interest were identified.

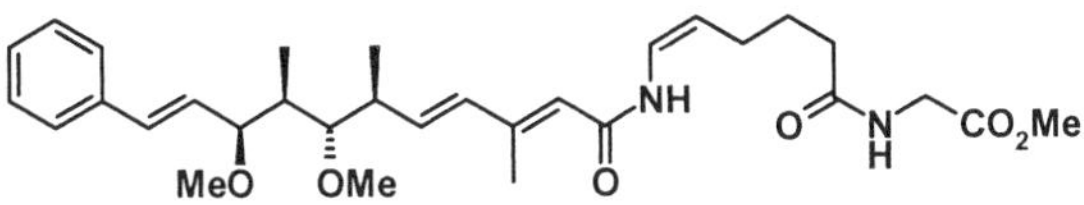

Crocacin D

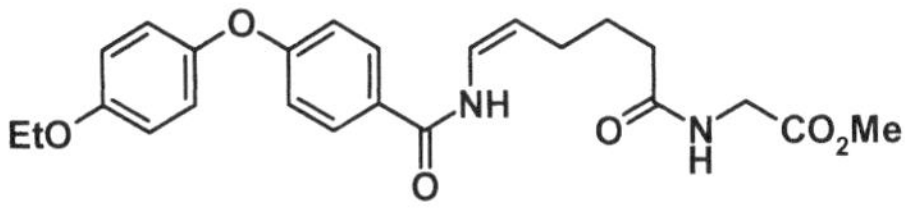

(4)

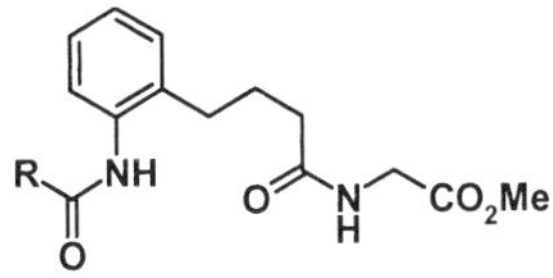

(5)

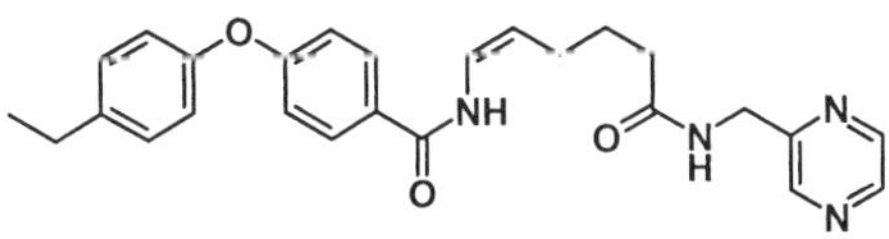

(6)

Figure 5 *Crocacin D and synthetic analogues*

3.5 The herbicidal triketones[13]

Mesotrione is an important post-emergence maize-selective herbicide, sold by Syngenta (Figure 6). It is a member of the triketone family of herbicides, compounds which act by inhibition of the enzyme hydroxyphenylpyruvate dioxygenase (HPPD) in plants, a mode of action which leads to a characteristic bleaching symptomology. It is acidic (pKa = 3.12 at 20°C) and has high water-solubility (15g/litre at pH 6.9 and 20°C), ideal properties for uptake and movement in plants. Importantly in the context of this paper, mesotrione is related to the natural product leptospermone, produced by the red bottle brush plant, *Callistemon citrinus*. Leptospermone is only a moderately active herbicide, but it contains the important triketone toxophore, and is an inhibitor of HPPD and gives bleaching symptomology when applied to plants.

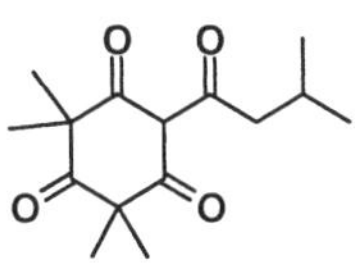

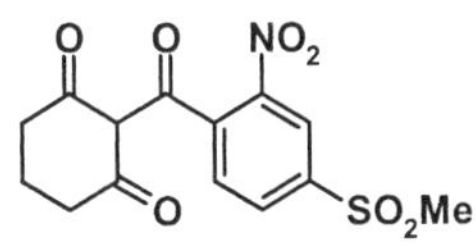

Leptospermone

Mesotrione

Figure 6 *The commercial synthetic herbicide mesotrione, related to the natural product leptospermone*

4 CONCLUSIONS AND OUTLOOK

Programmes of synthesis based on natural product leads often advance only slowly and are uncertain to deliver commercial products for crop protection. Many man-years of work may be invested without finding the single compound in which all the required properties are combined. Nevertheless, when successful, these projects lead to highly profitable products, and natural products remain an inspiration to chemists as an important source of novel structures and modes of action.

References and notes

1 Phillips McDougall – AgriService (Midlothian, Scotland), Industry Overview of the 2007 Market, published May 2008. Note that the figure quoted for the value of the world market for crop protection chemicals is the value of the sales of agrochemical products for crop use, including forestry and plantation crops, but does not include the sales of these products into non-crop situations.

2 C.A. Henrick, 'Pyrethroids' in *Agrochemicals from Natural Products*, ed., C.R.A. Godfrey, Marcel Dekker, New York, 1995, Chapter 2, pp. 63-145.

3 Y. Katsuda, *Pestic. Sci.*, 1999, **55**, 775.

4 T. Pitterna, 'Chloride Channel Activators/New Natural Products (Avermectins and Milbemycins)' in *Modern Crop Protection Compounds*, eds., W. Krämer and U. Schirmer, Wiley-VCH, Weinheim, 2007, Vol. 3, Chapter 29.6, pp. 1069-1088.

5 I.B. Bryan, M.J. Rice, M.R. Bartley, A.R. Jutsum and G. Pastushok, 'A Monic Acid Derivative: Evaluation as a Cereal Herbicide' in *Brighton Crop Protection Conf. - Weeds*, 1995, Vol. 2, pp. 725-730.

6 L. Pridzun, F. Sasse and H. Reichenbach, 'Inhibition of Fungal Acetyl-CoA Carboxylase: a Novel Target Discovered with the Myxobacterial Compound Soraphen' in *Antifungal Agents: Discovery and Mode of Action* (Papers presented at the conference The Discovery and Mode of Action of Antifungal Agents, Grantham, UK, 11-12 May 1995), eds., G.K. Dixon, L.G. Copping and D.W. Hollomon, Bios Scientific Publishers, Oxford, 1995, Chapter 9, pp. 99-109.

7 P. Bellini, J.M. Clough and G. Hatter, *Pestic. Sci.*, 1999, **55**, 491.

8 R.J. Lind, D.T. Greenhow, J. Blythe, J. Goodchild, E. Hirst, S.J. Dunbar and F.G.P. Earley, 'Cyanotropanes: Novel Chemistry Interacting at the Insect Nicotinic Acetylcholine Receptor' in *BCPC Conference - Pests & Diseases*, British Crop Protection Council, 2002, Vol. 1, pp. 145-152.

9 K. Beautement, J.M. Clough, P.J. de Fraine and C.R.A. Godfrey, *Pestic. Sci.*, 1991, **31**, 499.

10 J.M. Clough and C.R.A. Godfrey, 'The Strobilurin Fungicides' in *Fungicidal Activity - Chemical and Biological Approaches to Plant Protection*, eds., D.H. Hutson and J. Miyamoto, Wiley Series in Agrochemicals and Plant Protection, 1998, Chapter 5, pp. 109-148.

11 H. Sauter, W. Steglich and T. Anke, *Angew. Chem. Int. Ed.*, 1999, **38**, 1328.

12 P.J. Crowley, C.R.A. Godfrey and R. Viner, 'The Crocacins A and D: Novel Natural Products as Leads for Agrochemicals' in *Synthesis and Chemistry of Agrochemicals VII*, eds., J.W. Lyga and G. Theodoridis, ACS Symposium Series No. 948, 2007, Chapter 8, pp. 93-103.

13 G. Mitchell, D.W. Bartlett, T.E.M. Fraser, T.R. Hawkes, D.C. Holt, J.K. Townson and R.A. Wichert, *Pest Manag. Sci.*, 2001, **57**, 120.

EXPLOITING PHYTOCHEMICAL DIVERSITY TO CONTROL PESTS OF AGRICULTURE AND VECTORS OF HUMAN AND ANIMAL DISEASE

A.M. Hooper,[1] Z.R. Khan,[2] B. Torto,[2] M.A. Birkett,[1] T. J. Bruce,[1] J.A. Pickett[1]

[1]Biological Chemistry Department, Centre for Sustainable Pest and Disease Management, Rothamsted Research, Harpenden, Herts. AL5 2JQ, UK. E-mail: tony.hooper@bbsrc.ac.uk
[2]International Centre of Insect Physiology and Ecology, P.O. Box 30772, Nairobi, Kenya

1 INTRODUCTION

Ecological interactions between pests of agriculture and insect vectors of human and animal disease are mediated to a great extent through semiochemicals, signal molecules that convey information between organisms of the same or different species.[1] Due to the cost and expertise required to exploit these interactions through laboratory-based chemical synthesis of the molecules involved, the prospect of using identified natural products is enhanced through biodiversity, either by the availability of advanced synthetic intermediates as plant products to complete the synthetic process, or by plants themselves generating these signals *in situ*.

Semiochemicals can be placed into two distinct classes: *Pheromones* control intraspecific interactions. These semiochemicals provide information regarding a range of behaviours including mate selection, aggregation, dispersal (alarm pheromones), oviposition and food sources (trail pheromones). *Allelochemicals* control interspecific interactions such as host and non-host identification by primary colonizers (plant or animal) and whether a food source is already exploited through tritrophic interactions (signals indicating the presence of primary colonizers).

This review provides examples by which semiochemical exploitation strategies have been pursued either through synthetic chemistry based on the renewable resource of phytochemical diversity or through *in situ* production of semiochemicals by plants themselves. The chemistry that mediates ecological interactions in all instances is based in the chemical space defined by available biochemical processes. Plants, therefore, are a resource for these chemistries and often generate these specific signals as part of their own chemical ecology. The examination of ecological interactions through ethnobotany and new research or screening of essential oils or metabolomes provides a renewable resource through which non-toxic control of pest species can be further developed.

2 PHEROMONES

2.1 Mosquito Oviposition Pheromone

Mosquitoes of the genus *Culex*, such as *Cx. quinquefasciatus* and *Cx. pipiens*, vector human diseases such as lymphatic filariasis, encephalitis and the West Nile virus. Oviposition in these insects is influenced by an oviposition pheromone that is produced by egg-rafts as they near maturation, providing gravid females with information that the site is suitable for egg development.[2] The pheromone (5*R*,6*S*)-6-acetoxy-5-hexadecanolide can be synthesised by dihydroxylation and subsequent lactonisation of Δ5-hexadecanoic acid, a fatty acid of plant origin. Although it is rare to find this fatty acid in large quantities, *Kochia scoparia* (Chenopodiaceae), commonly called the burning bush, is a suitable plant source and its crude essential oil can be converted to material containing active pheromone using recyclable catalytic chemistry based on osmium tetraoxide.[3]

2.2 Aphid Sex and Alarm Pheromones

2.2.1 Aphid Sex Pheromones. Females of aphids in the subfamily Aphidinae call to alate males to mate using sex pheromones. The sex pheromones of this family are based on the structure of nepetalactone. By using different isomers of nepetalactone and its reduced form, nepetalactol, and in different blends, species-specific mating is achieved and not confused by the lack of structural diversity in the sex pheromone chemistry.[4] Nepetalactone can be synthesized by laboratory-based chemistry, but the high number of chiral centres and stereochemical requirements makes the process expensive. The plant *Nepeta cataria* (Lamiaceae), commonly called catnip, possesses (4a*S*,7*S*,7a*R*)-nepetalactone in its essential oil at 90% purity and as one enantiomer. Simple borohydride reduction generates its lactol (1*R*,4a*S*,7*S*,7a*R*)-nepetalactol and thus a synthetic blend of the pheromone can be generated cheaply.[5] In addition, the stereochemical isomer (1*RS*,4a*R*,7*S*,7a*S*)-nepetalactol, the sex pheromone of *P. humuli*, can also be isolated in good yield but from another species, *N. racemosa*. As well as being able to affect the behaviour of aphid pests, these semiochemicals are used by predators and parasitoids of pest aphids as foraging cues and thus can be used to attract them as part of integrated pest management strategies for aphid control.[6]

2.2.2 Aphid Alarm Pheromone. The alarm pheromone of aphids in the subfamily Aphidinae is (*E*)-β-farnesene.[4] The pheromone can be synthesized easily, but it is unstable under field conditions and therefore has limited potential for application. Although many plants produce this volatile chemical constitutively in their essential oil, aphids can differentiate between plant-produced material and aphid-produced material by the presence of other plant-derived terpenes, particularly (*E*)-caryophyllene, in the volatile blend.

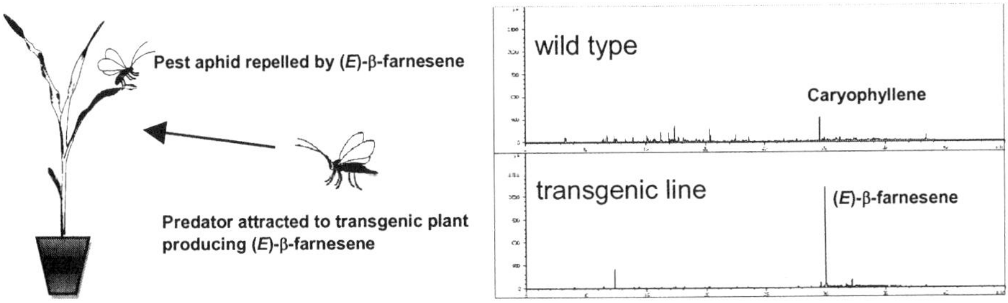

Figure 1 *Transgenic signaling;* Arabidopsis thaliana *emitting (*E*)-β-farnesene*

It has been possible to clone the (*E*)-β-farnesene synthase gene from *Mentha* x *piperita* (Lamiaceae) into *A. thaliana* (Brassicaceae) so that it is able to catalyse the conversion of native farnesyl pyrophosphate to (*E*)-β-farnesene, causing a change in the volatile chemistry (Figure 1).[7] This change is enough to reduce the colonization of the pest aphid *M. persicae* and increase foraging by the parasitoid *Aphidius ervi*. Therefore, this strategy could play a role in integrated pest management of crops affected by aphids if transgenic crops are created using this gene. The presence of most of the biosynthetic pathway in the plant allows the metabolome to be altered by introduction of a single gene to enable, in this instance, aphid control (see section **3.3**, *C*-glycoyltransferase activity for *Striga* control).

2.3 Sandfly Sex Pheromone

The sandfly *Lutzomyia longipalpis* is the vector of the causative agent of visceral leishmaniasis in the New World. Female *L. longipalpis* are highly attracted to leks of males that release a sex pheromone, so by attracting females to traps containing the sex pheromone lures, there is potential to reduce infection rates in humans. In the Lapinha region of Brazil (Minas Gerais State), the sex pheromone has been identified as (*S*)-9-methylgermacrene.[8] Whilst impressive total syntheses of this material have been achieved, the large number of chemical tranformations required make this an unsuitable method to obtain sufficient quantities of the pheromone for field trials, or for the transfer of production technology to areas where the pheromone needs to be deployed. The cranesbill, *Geranium macrorrhizum* (Geraniaceae), produces an essential oil (zdravet's oil) comprising 60% germacrone. With the complexity of 10-membered ring formation and the regio- and stereochemical complexity of three double bonds already in place, synthesis of the racemic pheromone can be completed in four steps.[9] The availability of this starting material has now allowed sandfly sex pheromone to be economically produced and supplied to researchers in the field.

3 ALLELOCHEMICALS

3.1 Antifeedants

Polygonum hydropiper (Polygonaceae), the marshpepper knotweed, produces polygodial in its essential oil. The compound is not toxic but plays an important part in the plant ecology as an antifeedant. In field trials, barley sprayed with polygodial was protected from the transmission of barley yellow dwarf virus as stylus probing by the aphid vector was attenuated by the antifeedant.[10] Other non-toxic antifeedants from *Ajuga remota* (Lamiaceae) (Ajugarin 1 and 2), though not active against aphids, were potent antifeedants for lepidopteran and coleopteran pests such as *Plutella xylostella* (diamondback moth), *Phaedon cochleariae* (mustard beetle) and *Leptinotarsa decemlineata* (Colorado potato beetle).[11] Antifeedant activity, as well as insecticidal activity, is an important feature of plant chemical defense, and phytochemical diversity in plant defense is a potent source of lead compounds for exploitation in crop and livestock protection.

3.2 Volatile non-host chemistry and tritrophic interactions

3.2.1 Haematophagous Insects. Mosquito repellents, such as DEET (*N,N*-diethyl-*m*-toluamide) have long been used in efforts to protect humans from these and other biting insects. Recent work on *Culex* spp. has suggested DEET works by an olfactory mechanism, stimulating the receptors that would normally respond to plant-derived odors such as eucalyptol, thus conveying the message of 'non-host'.[12] In species of *Anopheles*, *Aedes* and *Culicoides*, compounds have been isolated from humans that are relatively unattractive to these insects and these have been shown to mask the attraction of other humans giving protection from attack.[13] In East Africa, the gum of *Commiphora* spp. (Burseraceae) is used as a protection measure for cattle against ixodid tick pest populations. The biological activity of these gums has been analysed and shown to be dependant on the hydrocarbon content, an extract rich in sesquiterpenes such as δ-elemene, β-bourbonene and (+)-germacrene D. Material from *C. holtziana* (gum haggar) is rich in these compounds and is more effective than *C. myrrha* (gum myrrh).[14] The tsetse fly, *Glossina morsitans morsitans*, is a pest of cattle in the region but does not attack the Savannah water buck, *Kobus defassa*, again due to specific host and non-host semiochemicals which may be obtained through the diversity of plant secondary metabolism.[15]

3.2.2 Plant-insect Interactions. Maize, *Zea mays* (Poaceae), grown in small-holder farms in East Africa, is subject to attack from stem-borers, such as the caterpillar larvae of *Chilo partellus* and *Buseola fusca*. An intercropping and trap-cropping 'push-pull technology' has been developed which protects maize from the stem-borer using volatile semiochemicals.[16] Trap crops, such as Sudan grass, *Sorghum sudanense* (Poaceae), and Napier grass, *Pennisetum purpureum* (Poaceae), are planted around the maize and produce volatile semiochemicals, including naphthalene, linalool and octanal, that signal to the moth that it is a more suitable host-plant than maize and 'pull' the moths away from the maize inside the plot. At the same time, a repellent intercrop such as molasses grass, *Melinis minutiflora* (Poaceae), or green leaf, *Desmodium uncinatum* (Fabaceae), 'pushes' the moths away from the maize by emitting non-host volatile signals such as (*E*)-ocimene and (β)-caryophyllene.[17] In addition, molasses grass produces a nonatriene, (*E*)-4,8-dimethyl-1,3,7-nonatriene, a signal for herbivore damage that not only repels the egg-laying moth but attracts parasitoids such as *Cotesia sesamiae*, that subsequently search for the pest caterpillars to predate on.[18]

3.2.3 Plant-plant Interactions. One volatile component of plants distressed by herbivore damage can be the plant-signal chemical *cis*-jasmone. Treatment of wheat plants, *Triticum aestivum* (Poaceae), with *cis*-jasmone induces secondary metabolism that reduces settlement of the aphid pest *Sitobion avenae* and increases the foraging of *Aphidius ervi* parasitoids.[19] The signal up-regulates hydroxamic acids (DIMBOA, HBOA) and phenolic acid concentrations in the plant.[20]

3.3 Non-volatile allelopathy and *Striga* control

In the 'push-pull technology' developed to protect maize from stem-borers (see section *3.2.2*), the intercrop *D. uncinatum* was shown to prevent parasitism of maize by parasitic weeds from the genus *Striga* (Scrophulariaceae).[16] Subsequent work demonstrated this to be an allelopathic mechanism and biologically active compounds have been isolated from the *Desmodium* root exudate,[21] including isoschaftoside, 6-*C*-α-L-arabinopyranosyl-8-*C*-β-D-glucopyranosylapigenin.[22] The biosynthetic pathway to flavone production is ubiquitous in the plant kingdom, although the enzymic *C*-glycosylation required to biosynthesise this allelochemical is not. Studies are under way to screen other plants, especially edible legume crops, for metabolic pathways that could be altered to produce isoschaftoside and also to characterize the *C*-glycosyltransferase activity that is required for future breeding, screening or genetic manipulation.

4 CONCLUSION

This paper describes the diversity of plant secondary metabolism in the context of exploiting it for protection against pest organisms. Plants can be used either as sources of renewable fine chemicals for use in chemical synthesis to produce semiochemicals or as primary producers of these chemicals for extraction and isolation from essential oils. Alternatively, plants can be used in the field for *in situ* production of semiochemicals for combating pests of arable crops. Plant secondary metabolism can be altered to our benefit by breeding or genetic manipulation. Furthermore, as they too are organisms in the ecological environment, their secondary metabolism can also be altered to our benefit by semiochemicals.

References

1 A.M. Hooper and J.A. Pickett. *Semiochemistry* in *Encyclopedia of Supramolecular Chemistry*, ed. J. L. Atwood, Marcel Dekker, New York, 2004, p 1270.
2 L.E.G. Mboera, W. Takken, K.Y. Mdira and J.A. Pickett, *J. Med Entomol.*, 2000, **37**, 172.
3 T.O. Olagbemiro, M.A. Birkett and J.A. Pickett, *J. Ag. Food Chem.*, 1999, **47**, 3411.
4 J.A. Pickett, L.J. Wadhams and C.M. Woodcock, *Annu. Rev. Entomol.*, 1992, **37**, 67.
5 M.A. Birkett and J.A. Pickett, *Phytochemistry*, 2003, **62**, 651.
6 W. Powell and J.A. Pickett, *Pest Manag. Sci.*, 2003, **59**, 149.
7 M.H. Beale, M.A. Birkett, T.J. Bruce, K. Chamberlain, L.M. Field, A.K. Huttley, J.L. Martin, R. Parker, A.L. Phillips, J.A. Pickett, I.M. Prosser, P.R. Shewry, L.E. Smart, L.J. Wadhams, C.M. Woodcock and Y. Zhang, *P. Natl. Acad. Sci. USA*, 2006, **103**, 10509.

8 J.G.C. Hamilton, A.M. Hooper, H.C. Ibbotson, K. Mori, S.E. Muto and J.A. Pickett, *Chem. Commun.*, 1999, **23**, 2335.
9 A.M. Hooper, J.B. Farcet, N.P. Mulholland and J.A. Pickett, *Green Chem.*, 2006, **8**, 513.
10 G.W. Dawson, D.C. Griffiths, J.A. Pickett, R.T. Plumb, C.M. Woodcock and Z.N. Zhang, *Pestic. Sci.*, 1988, **22**, 17.
11 D.C. Griffiths, A. Hassanali, L.A. Merritt, A. Mudd, J.A. Pickett, L.E. Smart, L.J. Wadhams and C.M. Woodcock, in *Brighton Crop Protection Conference. Pests and Diseases*, BCPC, Surrey, 1988, p 1041.
12 J.A. Pickett, M.A. Birkett and J.G. Logan, *P. Natl. Acad. Sci. USA*, 2008, **105**, 13195.
13 J.G. Logan, M.A. Birkett, S.J. Clark, S. Powers, N.J. Seal, L.J. Wadhams, A.J. Mordue and J.A. Pickett, *J. Chem. Ecol.*, 2008, **34**, 308.
14 M.A. Birkett, S. Al Abassi, T. Krober, K. Chamberlain, A.M. Hooper, P.M. Guerin, J. Pettersson, J.A. Pickett, R. Slade and L.J. Wadhams, *Phytochemistry*, 2008, **69**, 1710.
15 N.K. Gikonyo, A. Hassanali, P.G.N. Niagi and R.K. Saini, *J. Chem. Ecol.*, 2003, **29**, 2331.
16 Z.R. Khan, J.A. Pickett, L.J. Wadhams, A. Hassanali and C.A.O. Midega, *Crop Prot.*, 2006, **25**, 989
17 M.A. Birkett, K. Chamberlain, Z.R. Khan, J.A. Pickett, T. Toshova, L.J. Wadhams and C.M. Woodcock, *J. Chem Ecol.*, 2006, **32**, 2475.
18 Z.R. Khan, K. AmpongNyarko, P. Chiliswa, A. Hassanail, S. Kimani, W. Lwande, W.A. Overholt, J.A. Pickett, L.E. Smart, L.J. Wadhams and C.M. Woodcock, *Nature*, 1997, **388**, 631.
19 T.J.A. Bruce, J.L. Martin, J.A. Pickett, B.J. Pye, L.E. Smart and L.J. Wadhams, *Pest Manag. Sci.*, 2003, **59**, 1031.
20 M.C.B. Moraes, M.A. Birkett, R. Gordon-Weekes, L.E. Smart, J.L. Martin, B.J. Pye, R. Bromilow and J.A. Pickett, *Phytochemistry*, 2008, **69**, 9.
21 M.K. Tsanuo, A. Hassanali, A.M. Hooper, Z. Khan, F. Kaberia, J.A. Pickett and L.J. Wadhams, *Phytochemistry*, 2003, **64**, 265.
22 Z.R. Khan, J.A. Pickett, A. Hassanali, A.M. Hooper, C.A.O. Midega, *Weed Res.*, 2008, **48**, 302.

ROOT CHEMICALS COULD OFFER OPPORTUNITIES FOR BREEDING FOR SWEET POTATO RESISTANCE TO THE WEEVIL *Cylas puncticollis* Boheman (COLEOPTERA: APIONIDAE)

H. Muyinza,[1,3] P.C. Stevenson,[2,4] H. Talwana,[3] D.R. Hall,[2] D. I. Farman[2] and R.O.M. Mwanga.[3]

[1]National Crops Resources Research Institute, Namulonge, P.O. Box 7084 Kampala, Uganda. [2]Natural Resources Institute, University of Greenwich, Chatham Maritime, Kent, ME4 4TB, UK. [3]Department of Crop Science, Makerere University, P.O. Box 7062, Kampala, Uganda. [4]Royal Botanic Gardens, Kew, Richmond, Surrey, TYW9 3AB, UK.

1 INTRODUCTION

Sweet potato (*Ipomoea batatas* Lam.; Family Convulvulaceae) is the 7th most important food crop globally with an estimated annual production of 135 million metric tonnes.[1] Uganda is the second largest producer of sweet potato in the world after China and the largest producer in Africa. The annual national production is estimated at 2.60 million metric tonnes, and the per capita consumption is about 110 kg.[2] Sweet potato is cultivated as an annual crop, propagated from cuttings of the first 30-45 cm of the apical stems. There is great variation among the sweet potato varieties in terms of foliage, root characteristics, food value and consumer acceptance and maturity period.[2]

In sub-Saharan Africa, sweet potato is a staple food for a large proportion of the population. It is easy to propagate and is highly adaptable, tolerant to high temperatures, low fertility soils and to drought. These characteristics make the crop ideal in mitigating famine and ensuring food security for numerous communities in many parts of Africa. Sweet potato roots are rich in ß-carotene, thiamine and vitamin C.[2] Comparatively, fresh roots contain 1.8-16 mg/100 g and dried roots contain approximately 7.8 mg/100 g of ß-carotene.

The sweet potato weevils *Cylas puncticollis* Boheman and *Cylas brunneus* Fabricius are the most important pests of sweet potatoes in Africa. Damage due to these pests have been reported to be as high as 60-97%.[3,4] Females of the two *Cylas* species lay yellowish eggs in cavities excavated in storage roots or stem bases and seal the cavity with faecal plugs.[5] On hatching, larvae bore into the sweet potato tissues and undergo three instars, during which they feed internally and tunnel extensively. The developing larvae then tunnel while feeding within the root or vine. These are the most destructive stages of the pest, and as a result of their feeding, the market value of the roots is significantly lowered.[4] Soon after leaving the pupal stage the adult emerges.[6]

Management options against the weevils have included use of cultural methods including hilling up to reduce weevil access to developing roots and planting of weevil-free materials.[5] These methods, however, have not provided sufficient protection, leading to the need for more effective management options. There are variations in levels of resistance to *Cylas* species among varieties of sweet potato.[11] However, in some instances the relative resistance observed among some varieties to the weevil has been attributed to be due to an 'escape' mechanism known as pseudo-resistance. This involves early maturing varieties which are harvested before the weevil population builds up. Additionally, deep-rooted varieties escape weevil damage because their roots are less accessible for females to lay eggs.[5] Active host plant resistance, on the other hand, could offer an effective, user-friendly management option against the weevil.

Sweet potato root and vine latex has an effect on the feeding and oviposition of *C. formicarius*.[7] Further, other workers reported differences in the feeding and oviposition preferences on the periderm and inner cores of storage roots by *C. formicarius*.[4] This could imply that there are differences in the acceptability of substrates by the *Cylas* spp. Indeed, certain chemicals at the root surface possibly acted as oviposition stimulants and these seemed to depend not only on the variety but also on cultivation conditions.[11-13,16] However, there is a need to investigate the effect of chemical root compounds in the resistant varieties on the development of *C. puncticollis* and *C. brunneus*. This study thus aimed at investigating the relationship between root chemical composition in resistant New Kawogo as compared to susceptible Tanzania variety and levels of resistance to *Cylas puncticollis* species. The identified resistance compounds would then be isolated and developed for incorporation into sweet potato for resistance breeding.

2 MATERIALS AND METHODS

2.1 Investigating the Variation in Oviposition and Feeding on Roots of Tanzania and New Kawogo

Undamaged roots of New Kawogo and Tanzania varieties were washed slightly with water and air dried. Plastic cylinders measuring 8 mm (diameter) x and 12 mm (length) with openings at each end were then used to cage weevils in a transparent arena on the roots. Five weevils were placed on each of the root sites. To enable the weevils to choose between the two varieties, different varieties were placed at each open end of the cylinder and then sealed and held in place using moist pliable clay. The bio-assay was left to stand for 24 hours, and then the numbers of droppings, eggs laid and feeding punctures of each of the arenas were determined.

2.2 Chemistry of New Kawogo and Tanzania Root Latex

2.2.1 Root latex collection and root surface extraction, and the use of LC-MS and HPLC to identify root chemicals. Freshly harvested roots of Tanzania and New Kawogo varieties were broken in two by hand, which ensures maximum flow of latex, as opposed to cutting using a sharp knife which seems to inhibit the flow. The latex was immediately collected from the broken exposed surface with a melting point/capillary tube. An aliquot of the latex was transferred to a vial, weighed, dissolved in methanol and stored at 3 ± 3 ^{0}C. Root latex from New Kawogo and Tanzania was extracted in methanol and the extracts were compared by HPLC analysis. Whole undamaged roots of New Kawogo were submerged in 1 litre of hexane for 1 minute. The extract was filtered through Whatman Grade 1 paper and evaporated to dryness under reduced pressure on a rotary evaporator and then redissolved in methanol. Liquid chromatography-mass spectrometry and HPLC were done using the Waters system with 600E quaternary pump, 717 autosampler and 996 diode array detector. Chromatographic separation was performed on a 150 mm x 4.6 mm i.d. (5 mm particle size) Zorbax Eclipse C18 column using a linear mobile phase gradient of 0.94 ml/min flow rate with 5% acetic acid (A); MeOH (B); and 18:1:1 MeOH: water: acetic acid (C). Initial conditions were 80% A, 0% B and 20% C, changing to 80% B and 20% C at T = 20 minutes and finally to 100% B at 21 minutes, this final solvent being maintained for 41 minutes. The injection volume was 5 μl and data analysis was performed using Millenium software.

LC-MS was carried out on a Thermo-Finnigan LC/MS/MS system consisting of a 'Surveyor' autosampling LC system interfaced to a LCQ Classic quadrupole ion trap mass spectrometer. The ion trap MS was fitted with an Atmospheric Pressure Chemical Ionisation (APCI) source operated under standard conditions; i.e. vaporiser temperature 450 °C, needle current 5 mA, heated capillary temperature 150 °C, sheath and auxiliary nitrogen gas pressure 80 and 20 psi, and the source voltages tuned for the optimal transmission of protonated rutin. The ion trap was set to monitor ions from m/z 125-1200 with collision energy of 45%. Chromatographic separation was performed on a 150 mm x 4.6 mm i.d. (5 μm) Phenomenex Luna C18 column using a linear mobile phase gradient of 1 ml/minute flow rate, and the following solvents and solvent mixtures: water (A); MeOH (B); 5% HOAc in MeOH (C). A mixture of 80% A and 20% C was used initially, switching after 25 minutes to a mixture of 80% B and 20% C. Injection volumes were 10 μl and data analysis was performed using Xcalibur 1.2 software.

Fragmentation in APCI-MS(+) was characterised by the molecular ion of each compound with two fragments representing the substituted cinnamate moiety (e.g., coumarate and coumaroyl). For example, the molecular ion for hexadecyl *p*-coumarate (**2**) was *m/z* 389 $[M + H]^+$ with fragments at *m/z* 165 (coumaric acid) and *m/z* 147 (coumaroyl), consistent with loss of $-C_{16}H_{33}$ and $[O-C_{16}H_{33}]^+$ respectively. The total quantities of the identified compounds were evaluated from the chromatograms using the area under each spectrum and compared to a standard. The compounds were also artificially synthesised and used for incorporation into diet bio-assays.

2.2.2 Synthesis of hexadecyl esters of caffeic and p-coumaric acids. Compounds **1** and **2** (Figure 2) were prepared in quantitative yields from the corresponding acids. The hydroxyl group(s) were acetylated to give the acetoxy acids and these were then allowed to react with the appropriate alcohol, R_1OH, using N,N′-dicyclohexylcarbodiimide catalysed by N,N-dimethylaminopyridine in dichloromethane.[14] The acetyl groups were then removed with potassium carbonate in methanol to give the esters which precipitated out on acidification. The products were recrystallised from ethyl acetate/petroleum spirit and characterised by IR, LC-MS and

NMR spectroscopy. Up to 1 g of hexadecyl caffeate and hexadecyl *p*-coumarate were synthesised for bioassays and analysis.

2.3 Effect of Synthetic Hexadecyl *p*-Coumarate and Hexadecyl Caffeate on Larval Development

A diet developed to rear larval stages of *C. puncticollis* and C. *brunneus* to adulthood was used to investigate the biological activity of compounds **1** and **2**.[15] The compounds were incorporated into diets as solutions in ethanol. Control diets were treated with the same volume of ethanol only, to demonstrate the absence of an adverse effect of this component of the diet on the development of the larvae. To 90 g of diet, **1** and **2** were added respectively at concentrations of 0.1, 0.01, 0.001 and 0.0001% w/w and mixed using a magnetic stirrer. The diet, while still hot, was then poured into sterile plastic Petri dishes with 7 replicates per treatment. After the diet had cooled and set, 5 small burrows were excavated in the diet surface using a plastic spatula to scoop out a small piece of diet. When larvae were placed in the burrow the displaced diet was replaced on top to provide protection from desiccation.

Second instar *C. puncticollis* larvae (11 days old) were obtained by allowing adult mated weevils to oviposit on the roots for 24 hours. The age of the weevils was recorded from the first day of incubation. One larva was placed into each diet burrow with 5 burrows on each Petri dish. The Petri dishes were then covered with a disc of filter paper to absorb excess moisture from the diets and the lids were replaced and the bio-assay left to stand for 10 days. The percentage larval mortality was recorded.

3 RESULTS AND DISCUSSION

3.1 Variation in Mean Weevil Development, Oviposition and Feeding of *C. puncticollis* on Roots of Tanzania and New Kawogo

There were significantly more feeding punctures, droppings and eggs laid by insects feeding on root plugs of Tanzania compared with insects feeding on New Kawogo root plugs (Figure 1), indicating that an active and quantifiable mechanism of resistance was present in the roots of New Kawogo.

This result reciprocates deterrent effects reported by earlier workers[7] who showed that latex-coated root cores were less preferred for oviposition and that fewer feeding punctures were recorded in semi-artificial diets containing the latex. This suggests that the resistance of New Kawogo may be associated with components in the latex. The difference in the number of droppings reflects differences in consumption between adults feeding on the two varieties and suggests that fewer feeding punctures does indicate reduced consumption and thus damage.

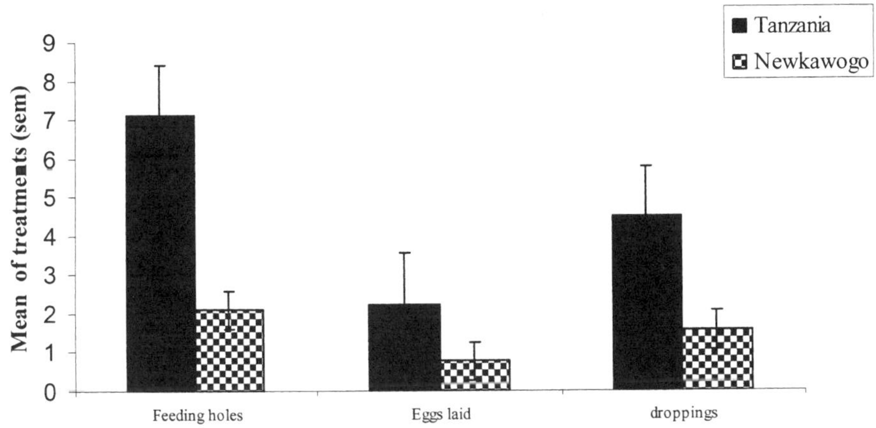

Figure 1 *Numbers of feeding holes, eggs laid and droppings of adult Cylas puncticollis on root plugs of two varieties of sweet potato - after 24 hours (n = 16), (sem = standard error of the mean)*

3.2 Chemistry of New Kawogo and Tanzania root latex

The four most abundant compounds in the root latex of the sweet potato roots were identified by comparison with standards synthesised in our laboratory as octadecyl and hexadecyl esters of caffeic and *p*-coumaric acids (**1** – **4,** Figure 2). The most abundant compounds in the latex of Tanzania were **2** and **4** with only a small amount of **1** in comparison to the amounts in New Kawogo (Figure 4).

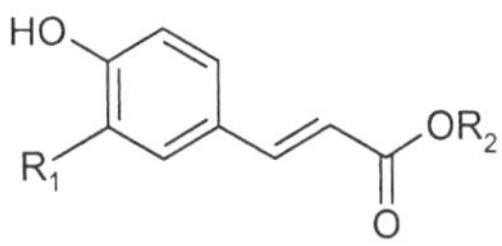

		R_1	R_2
1	Hexadecyl caffeate	OH	$-(CH_2)_{15}CH_3$
2	Hexadecyl *p*-coumarate	H	$-(CH_2)_{15}CH_3$
3	Octadecyl caffeate	OH	$-(CH_2)_{17}CH_3$
4	Octadecyl *p*-coumarate	H	$-(CH_2)_{17}CH_3$

Figure 2 *Compounds extracted from root latex of the sweet potato cultivars New Kawogo and Tanzania*

The chemical profile of New Kawogo root latex, which is also presented in Figure 3, contained peaks for **1**, **2** and **4** but also contained **3**, tentatively identified as octadecyl caffeate. This compound appears to be present at much lower concentrations in the variety Tanzania. Compounds **1** and **3** have not previously been found in sweet potato and so are reported here for the first time. When the synthetic compound **2** (100% *E*-form) was left in solution for 3 days at room temperature a second compound appeared, eluting approximately 30 seconds earlier and with a similar mass spectrum and a UV spectrum consistent with the *Z*-form of coumaric acid. Thus it is presently proposed that the peak eluting in front of **4** in Figure 3 is the *Z*-form of **4** since this peak has an identical MS to **4** and a UV spectrum consistent with the *Z*-form of octadecyl *p*-coumarate.

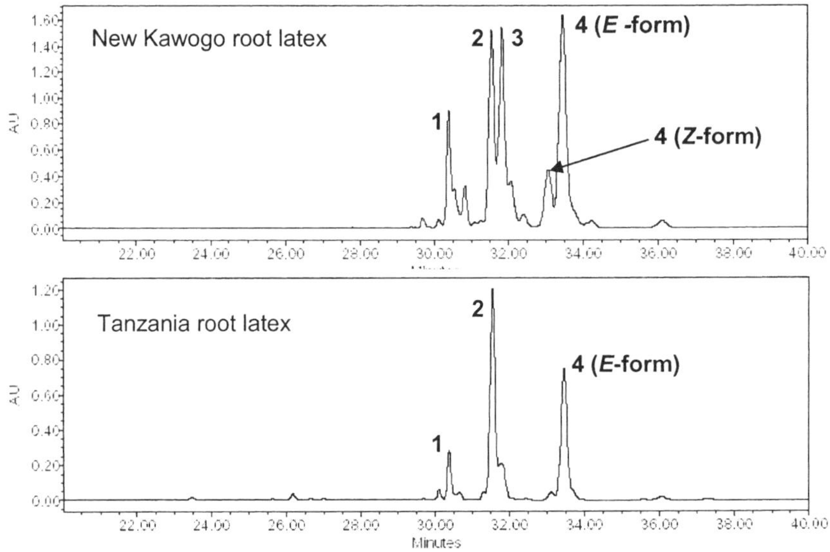

Figure 3 *HPLC chromatograms at 325 nm of New Kawogo and Tanzania Root latex (40 mg latex/ml) illustrating differences in chemical profile*

There was a significantly higher concentration of caffeic and coumaric acid esters in New Kawogo latex as compared to those found in the variety Tanzania (Figure 3). The incorporated compounds increased the mortality of 2nd instar larvae significantly (Figure 4). This effect was dose-dependent. This clearly shows that the compounds are active and affect the development of *Cylas* weevils in sweet potato, and could account for the differences in resistance among varieties.

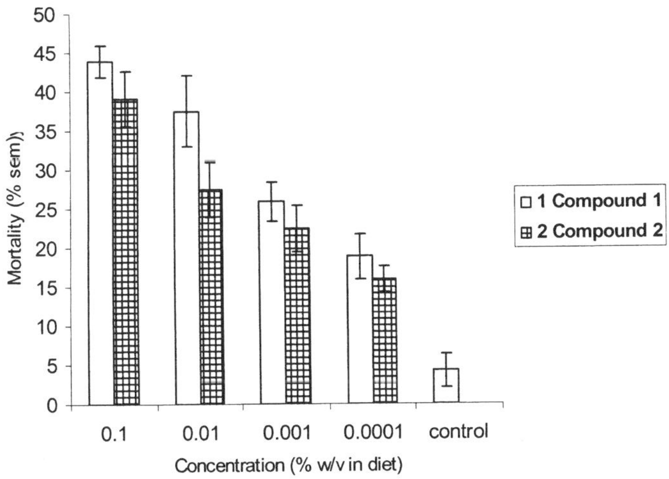

Figure 4 *Effect of incorporating compounds* ***1*** *and* ***2*** *into diets at different concentrations on the mortality % (±sem) of sweet potato weevil larvae*

Previous work on *C. formicarius* implicated the active role sweet potato latex might play on the feeding and oviposition by sweet potato weevils. All the parts of the sweet potato plant on which the weevil feeds, including the roots, leaves and vines, exude copious amounts of latex upon wounding.[7] Furthermore, young vegetative parts of sweet potato produced more latex and tend to show reduced weevil feeding damage than older and more mature portions of the vine. Latex incorporated into semi-artificial diets reduced the number of feeding punctures and when coated onto root surfaces resulted in lower oviposition. Earlier work had already shown that components of the latex included hexadecyl, octadecyl and eicosyl esters of coumaric acid,[16] suggesting that it is the presence of these compounds that is responsible for the deterrent effects on the feeding and oviposition of the adults.

This study has shown that the number of feeding holes by *C. puncticollis* differed significantly between the resistant and susceptible varieties. Since the two varieties differed quantitatively and qualitatively in the presence of octadecyl and hexadecyl esters of coumaric and caffeic acids, these chemical differences may well explain the observed behavioural responses of the insects, and this is in line with earlier findings.[7]

However, while larvae do reach adulthood on the artificial diet used in the bioassays reported here, the insects are still likely to be under greater stress than when feeding on whole roots. Thus the effects of the compounds may not exactly represent their effect when encountered in the living root. Nevertheless, the specific identification of **1** and **2** in New Kawogo and their observed effect on larval development are evidence that resistance in sweet potato is active and quantifiable and thus useful in breeding for varieties that are resistant to *C. puncticollis*. In addition, the higher levels of these compounds obtained from New Kawogo as compared to Tanzania strongly suggest that varieties with higher concentration of these compounds in the roots will be less damaged by *Cylas* spp.

The identification of all these esters on the root surface is also significant in the effort to identify resistance mechanisms in African sweet potato to *C. puncticollis*; particularly since they affect early development of larvae and may also affect egg laying behaviour. Thus breeding for varieties with higher quantities of hexadecylcaffeic and hexadecyl-*p*-coumaric acids offers a viable option in the effort to develop varieties with resistance against the weevil *C. puncticollis*. However, the findings in this study are in contrast to earlier reports on other resistant African sweet potato cultivars,[9,10] where it was postulated that resistance in sweet potato could have been due to the so-called pseudo-resistance described in the introduction.

4 CONCLUSION

The findings in this study clearly provide the proof that resistance in New Kawogo must be more than simply due to the escape mechanism of pseudo-resistance, and is active, quantifiable and potentially manageable for breeding purposes. The inheritance of the root latex esters needs to be studied in new crosses, and mapped in new populations.

References

1 http://faostat.fao.org/faostat/servlet. FAOSTAT Database results 2008.

2 J.A. Wolfe, Sweet potato: An Untapped Food Resource. Cambridge University Press Cambridge, U.K. 1992, p 643.

3 M.A. Mullen, *J. Agric. Entomol*, 1984, **1**, 227.

4 S.F. Nottingham, S.J. Kays, *Acta Horticulturae*, 2002, **583**, 155.

5 N.E.J.M. Smit, Integrated pest management for sweet potato in Eastern Africa. Grafish Service Centrum Van Gils B.V., Wageningen, 1997, p. 2-117.

6 CAB International, Crop Protection Compendium, Global Module, 2nd Edition. CAB International, Wallingford, UK, 2000.

7 E.S. Data, S.F. Nottingham and S.J. Kays, *J. Econ. Entomol.*, 1996, **89**, 544.

8 T.E. Stathers, D. Rees and D. Jeffries, Investigating the potential of cultivar differences in susceptibility to sweet potato weevil as a means of control. *Final Technical report.* Department for International Development (DFID) Crop Post-Harvest Programme. Natural Resources Institute 1999.

9 T.E. Stathers, D. Rees, S. Kabi, L. Mbiliny, N. Smit, H. Kiozya, S. Jeremiah, A. Nyango and D. Jeffries, *Int. Pest Manag.*, 2003, **49**, 131.

10 T.E. Stathers, D. Rees, A. Nyango, H. Kiozya, L. Mbilinyi, S. Jeremiah, S. Kabi and N. Smit, *Int. J. Pest Manag.*, 2003, **49**, 141.

11 K.C. Son, R.F. Severson, M.E. Snook and S.J. Kays, *HortScience*, 1991, **26**, 1305.

12 K.C. Son, R.F. Severson and S.J. Kays, *HortScience,* 1991, **26**, 1514.

13 K.C. Son, R.F. Severson and S.J. Kays, *HortScience,* 1991, **26**, 409.

14 B. Neises and W. Steglich, *Angewandte Chemie*, 1978, **17**, 522.

15 M. Ekobu. An artificial diet for screening *Bacillus thuringiensis* Cry proteins against sweet potato weevils, *Cylas puncticollis* and *Cylas brunneus*. MSc thesis, Faculty of Agriculture, Makerere University, 2006, p. 1-99.

16 M.E. Snook, E.S. Data and S.J. Kays, *J. Agric. Food Chem.*, 1994, **42**, 2589.

THE ROLE OF BIOTECHNOLOGY IN MEETING THE BIODIVERSITY CONSERVATION CHALLENGE OF AFRICA

A.B. Salifu

Council for Scientific & Industrial Research – Ghana (CSIR – Ghana), P.O. Box M.32, Accra, GHANA

1 INTRODUCTION

Since the early 1980s when biotechnology came into commercial application as an emerging technology, its impact on national economies has become more and more manifest. Biotechnology applications are becoming increasingly profound, especially with the developments in genomics, bioinformatics and genetic engineering.[1] What is of essence in the trends in biotechnology applications is the connection to the development challenges of specific sub-regions of the world. Africa is at the centre of the debate on how to stimulate growth, improve the lives of the people in their respective societies and ensure sustainable development of its environment.

Indeed Africa's present conditions depict a continent trapped in a state of development mire. Poverty incidence averages at about 40% living below the income level of US$1 a day. Sierra Leone, Liberia, Chad, Guinea Bissau and Niger have poverty incidence of about 70% below the poverty line of US$1 income a day. Outside South Africa and a few other countries, industrialization, by which the countries integrate into the world economy and compete for a fair share of the global resources, is rather minimal. Much of its industrialization is based on the extraction of its natural resources with serious consequences to its environment. For example, the mining of gold and diamonds in Ghana, the drilling of oil in Nigeria and Angola, and the logging and exports of timber from the Congo are all posing great challenges for environmental management in Africa. All this constitutes a challenge for the conservation of biodiversity.

So what role can biotechnology play in meeting the challenges associated with the conservation of biodiversity? How are the current levels of development in the adoption and application of biotechnology in the different countries able to deal with the challenges? What are the options for enhancing the role of biotechnology in the conservation of biodiversity? This paper discusses these and other related issues.

2 THE BIODIVERSITY CONSERVATION CHALLENGE

Biodiversity has been at the heart of socio-economic sustainability for the people. For thousands of years, village communities have developed their own concepts and practices in the exploitation of their respective biodiversity and its management. Biodiversity had become, and continues to be, an integral part of their lives.

With the advent of the modern state human societies have come the challenges of keeping to the time-tested practices in exploiting the resources of the land and water ecosystems. By tradition, socio-economic and cultural practices, such as clearing farmlands, harvesting wood as domestic source of energy, hunting for meat and extracting inputs for building dwelling places, have put significant strain on biodiversity. But the greater strain has come from modern society, whose perceived acts of modernity continue to impact greatly on traditional conservation practices. In Africa, the construction of hydro-electric dams on the Volta, the Nile, the Niger, the Congo and several other rivers has transformed ecosystems and their biodiversity. Further, the oftentimes unplanned setting up of industries, the urbanization and expansion of cities, and the construction of roads, have brought disruptions to Africa's biodiversity resources. About 50% of the people in Africa are projected to be living in urban areas by 2020. The indications are clear that, should the trend continue, the rich biodiversity resources of Africa would evaporate. Biodiversity conservation therefore becomes an urgent and overarching imperative.

The challenge for biodiversity conservation also comes in some other ways. Invasive alien species (IAS) are the second greatest cause of biodiversity loss on planet earth, next to habitat destruction that poses even greater threats. The threat of IAS is one that African countries are now grappling with. Tropical American floating water plants, like water fern, water lettuce and water hyacinth, already form thick mats on water bodies in many African countries, reducing light and oxygen, affecting biodiversity, fishing, water availability, power generation and navigation. The threat of IAS and the barriers constraining their management in Africa were the motivations behind the development of the project "*Removing Barriers to Invasive Plant Management in Africa*", established by the United Nations Environmental Program (UNEP) and the Global Environment Facility (GEF). The participating countries are Ethiopia, Ghana, Uganda and Zambia with project coordination being undertaken by CABI and IUCN. In Ghana, the project is being coordinated by the Council for Scientific and Industrial Research (CSIR).

Generally, many of the problems constituting the challenge in biodiversity conservation have institutional, human and technological dimensions. Whilst attempts have been made to address the problems from these dimensions, it is important to consider how to do this holistically to ensure synergy.

3 BIOTECHNOLOGY

Modern biotechnology offers an immense new knowledge and vastly greater technological options in working with genes and genomes within and across living systems. The technology enables the production of new plant varieties with precision and scope unimaginable just a few decades ago. However, the extent to which modern biotechnology is effectively applied to meet particular challenges depends on procedures developed by national institutions and organizations responsible for developing them. More so, the effectiveness of biotechnology depends on the links formed between the institutions and those who need to enact the procedures. In this regard, the functions of key actors such as scientists, policy-makers, farmers, industrialists and regulators are crucial.

The development of biotechnology in selected African countries has been surveyed, and it was concluded that, whilst significant progress has been made in efforts to apply biotechnology to specific problems in agriculture, there is a gap between what pertains in African countries and what is practised elsewhere. Nigeria, Senegal, Ghana, Cameroun, Uganda and Kenya are examples of countries that are building institutions and developing capacity for the application of biotechnology. However, in these countries, this work has not reached the stage where local scientists and their institutions can transform the genetic resources as for example in farming.

Such transformation of genetic resources has become a key indicator in driving agricultural performance in several countries. For example, whilst the cultivation of genetically modified (GM) crops has, in the past, been confined mainly to a few industrialized countries, notably the US, it is now becoming pervasive. In 2007, the global cultivation of GM crops increased to a total area of 143.7 million hectares, a 22% increase over the global area planted in 2006.[2] Countries such as China, India, Brazil and Argentina are fueling the spread in the farming of GM crops in a big way. In Africa, South Africa and Egypt are taking the lead in their cultivation, with field trials coming up in Kenya, Burkina Faso and a few other countries. The issue still remains that there is need for the development of national capacities to determine the direction of biotechnology application on the continent. Of course, the transformation of genetic resources in itself raises issues in relation to biodiversity conservation. However, the argument here is that, once the capacity exists, its exploitation can be determined to address specific national objectives, including implementing projects contributing to biodiversity conservation.

In this regard, the example of the Forest Research Institute of Ghana (FORIG) is instructive, and this, to some extent, is similar to what pertains in analogous research institutes in Nigeria, Kenya and other African countries. FORIG plays a major role in Ghana's Presidential Special Initiative in Forest Plantation Development. Ghana's National Forest Development Programme was launched in September 2000, aimed at producing 20,000 hectares of planted forest annually. The Institute developed methods for the collection, processing, testing, storage, and pre-germination treatment of seeds, and for the culturing of seedlings, of the following economically important forest species: Emire (*Terminalia ivoriensis*), Ofram (*Terminalia superba*), Kusia (*Nauclea didderichia*), Mahogany (*Khaya spp.*), Edinam (*Entandrophragma angolense*), Wawa (*Triplochiton scleroxylon*), Teak (*Tectona grandis*) and Pine (*Pinus spp.*). It is estimated that about 102,000 hectares of forest would be re-planted under the initiative.

FORIG has been able to establish a functional biotechnology laboratory equipped to undertake various procedures, including DNA amplification, electrophoresis, and tissue culture. Biotechnology tools are used to characterize the forest species listed above, and to produce high quality planting materials. Bulk analysis assists in the identification of DNA markers linked with desirable growth traits, particularly of teak. Where applicable, tissue culture has also been employed. A parallel initiative is analyzing genetic diversity within bamboo for identification of unrelated culms for the establishment of plantations.

The capacity created at FORIG has been used to train students from certain West African universities in biotechnological approaches for the conservation of forest plantations and related activities. Most students that have been trained, or are being trained, come from the Ghanaian Kwame Nkrumah University of Science and Technology, the University of Parakou in the Republic of Benin, or are researchers in other institutions. This training initiative and its widespread coverage perhaps signify that biotechnology has come to stay as a significant potential contributor to biodiversity conservation.

4 BIOSAFETY IN AFRICA

Even in the 1970s, when modern biotechnology was in the experimental stages, there was an awareness of the issue of safety. As the technology has developed, so has the concept of biosafety, which recognizes the need to minimize or eliminate the risks associated with the application of biotechnology. Efforts to institutionalize the assurance of safety came to a climax with the development of the Biosafety Protocol, also known as the Cartegena Protocol, which provides for public participation in the formulation of regulations associated with the application and commercialization of biotechnology. Thus Article 23 of the Protocol requires that the public is consulted in the decision-making process regarding living modified organisms (LMOs) and the results of the decision are also made public, while respecting confidential information.

Efforts to institutionalize biosafety in Ghana goes back to the 1990s, when a Biosafety Focal Point was instituted at the Biotechnology and Nuclear Agriculture Research Institute (BNARI). Since then, there have been efforts to set the mechanism in place for the formulation of the biosafety regime. Stakeholders such as the Institute of Biologists and the Science and Technology Policy Research Institute (STEPRI) of the Council for Scientific and Industrial Research (CSIR) held workshops and seminars and conducted studies to define the process for the formulation of the biosafety regime. The current effort at institutionalizing a biosafety regime came with the initiation of a project at the Ministry of Environment and Science, sponsored by the United Nations Environmental Program (UNEP) and the Global Environment Facility (GEF). Under the Project, Stakeholder Consultative Workshops were held to deliberate on the scope and content of the biosafety framework for Ghana. The culmination of the consultative process has been the drafting of a National Biosafety Bill and Guidelines for Biosafety, which has been ratified by Parliament.

The guidelines cover procedures that have to be followed for obtaining approval for modern biotechnological work for innovative applications. They embrace all work carried out in Ghana relating to gene manipulation and for employing nucleic acids for purposes including:

- the development of transgenic plants, animals and micro-organisms;
- the industrial manufacture of LMOs and products thereof; and

- the release of LMOs in closed systems, into the environment for field trials and for commercial purposes, either by public or private companies, as well as for export and import.

Public participation has also become important for the formulation of a National Biotechnology Policy for Ghana. The CSIR-STEPRI has also undertaken a project *"Towards the Systemic Evolution of a National Biotechnology Policy in Ghana"*, with the support of the International Development Research Centre (IDRC) of Canada, which harnessed inputs from the critical stakeholders to produce this National Biotechnology Policy. The Policy provides a vital framework for prioritizing and allocating resources to facilitate the development of biotechnology, and has been submitted to government for adoption.

Efforts have been made to institutionalize policies for biotechnology or biosafety in many African countries, such as Nigeria, Kenya, Uganda, Tanzania and South Africa, with the support of governments and international organizations such as the United States Agency for International Development (USAID). However, biotechnology or biosafety policies are not necessarily an end in themselves, and implementation is perhaps even more challenging than their formulation. A national capacity for biotechnology application has to be built over time and in a deliberate and planned manner within the framework of a national development plan.

Ghana has a *National Science and Technology Policy Statement Document*, which cabinet has adopted as the framework for science and technology development since 2000. However, since its adoption, it has virtually been lost in the national scheme of things and there is hardly any reference to it in the specific areas of science and technology programming and project implementation, nor in the planning for the broader and more generalized development process.

5 THE OPTIONS FOR BIOTECHNOLOGY FOR THE CONSERVATION OF BIODIVERSITY

There are a number of actions which various stakeholders, such as governments, NGOs, and public and private institutions, can carry out to enhance the role of biotechnology in the conservation of biodiversity. For example, one of the important initiatives in Ghana is the establishment of arboreta for the conservation of medicinal plants. Some of the medicinal plants are becoming endangered due to over-exploitation and deforestation. There are arboreta in the communities of Mamfe, Mampong, Ayikuma and Begoro, where medicinal plants such as *Cryptolepis sanguinolenta, Blighia sapida, Olax subscorpioidea* and *Khaya senegalensis* are being conserved. Some of these efforts can be facilitated with biotechnological applications for the characterization and cryo-preservation of the plants.

There are also policy and legislative initiatives which must be undertaken to ensure the conservation of Africa's biodiversity. Some of these only require the strengthening of existing legislations and policy instruments. For example, in Ghana there is the Concessions Ordinance of 1939, which empowers the Chief Conservator of Forests to ensure the protection, conservation, management, improvement, development and exploitation of all forest resource in concessions. These provisions for forests are buttressed by the Forest Ordinance, Cap 157 as amended by Cap 157A, and the Forest Amendment Act, 1957 No. 10. Complementary legislations have also been put in place to enforce the protection of the nation's forests. Examples are the Forest Protection Decree 1974 (NRCD 243), the Forest Plantation Development Act, Act 583, and the Forestry Regulatory Act, Act 547, amended to Act 617, termed as the Timber Resource Management (Amendment) Act 2002.

Laws are only as effective as the institutions which enforce them, and so there is a need to equip the respective institutions to carry out their legislated functions. It is not only a case of supplying the physical infrastructure, equipment and accessories, but there is also a need to put trained people in place, with access to new knowledge and the will to enforce the law. For example, to conform with international obligations on phyto-sanitation, a permit is always required to import any plant or part of a plant into African countries. Injurious plant species imported without permit should be confiscated and destroyed at the point of entry at a cost to the importer. As part of an arrangement to foster international collaboration, a health or phyto-sanitary certificate is issued to exporters of plants, plant parts or produce to other countries to avoid the spread of injurious pests, including potentially invasive alien species. However, the institutions responsible for ensuring phyto-sanitation and policing national borders need to be properly equipped and determined to carry out their responsibilities.

Several African countries have been assisted by international organizations to develop their national strategies for biodiversity conservation and to set up the institutional framework for their implementation. In addition to a National Focal Point, Ghana has also set up a National Biodiversity Committee as an advisory body to address the challenges under the Convention on Biological Diversity (CBD). To comply with the obligation of states under the CBD, the country has also prepared and published in 2002 its *National Biodiversity Strategy for Ghana* under the auspices of the Ministry of Environment and Science. The National Biodiversity Committee has also prepared the Action Plan for the implementation of the national strategy. The principles guiding the biodiversity strategy include the following:

- The conservation of ethics, including the inherent right to existence of all living forms, is deeply rooted in the indigenous socio-cultural values of Ghanaians.
- Biological resources are natural capital, and their conservation should be an investment that will yield benefits locally, nationally and globally for present and future generations.
- Benefits accruing from the use and conservation of biological diversity are to be shared equitably and fairly among all stakeholders. Benefits arising from the innovative use of traditional and customary knowledge of biological diversity will also be equitably shared with those from whom the knowledge has been acquired.
- The sustainable management of biological diversity is the responsibility of all sectors of society.

The implementation of such national strategies is the key to effective biodiversity conservation. Integral to the implementation process is the effort to build capacity to ensure that functions are effectively carried out.

6 CONCLUSIONS

Biotechnology has the potential to facilitate the conservation of biodiversity, but such potential needs to be exploited within appropriate policy and institutional regimes. The threat to biodiversity is manifest in Africa given the modernization and the demands for urbanization and industrialization. In recent years, invasive alien species have also become a problem in African countries. Given that most African countries have developed their national biodiversity strategies, there is a need for these countries to implement these strategies and to empower institutions to carry out their functions effectively.

References

1 A.E. Ochem (2006) "New Developments in Biotechnologies: Challenges for Africa", Paper presented at the Ministerial Conference of the ARCT Governing Council held in Accra, 4th – 6th May, 2006.
2 C. James (2007). Global Status of Commercialized Biotech/GM Crops: 2007, ISAAA Brief No. 37, ISAAA, Ithaca, N.Y., page 106.

FARMERS AS A TREASURE CHEST OF BIODIVERSITY - THE CASE OF SORGHUM [*Sorghum bicolor* (L.) moench] IN ETHIOPIA, A FIVE DECADE ANALYSIS

F. Mekbib

Haramaya University, P.O. Box 138, Dire Dawa, Ethiopia. **Norwegian Address:** Norwegian University of Life Sciences, Department of Plant and Environmental Sciences, P.O. Box 5503, N-1432, Aas, Norway. **Current corresponding address:** P.O. Box 485 code 1250, Addis Ababa, Ethiopia. E-mail: firewmekbib@yahoo.com

1 INTRODUCTION

At the world summit on sustainable development held in Johannesburg in South Africa in 2002, biodiversity was described as one of the five major areas for progress in achieving a sustainable future for humanity. The United Nations Convention on Biological Diversity[1] defines biodiversity as '*the variability among living organisms from all sources, including terrestrial, marine and the ecological complexes of which they are part.*' Agro-biodiversity encompasses the variety and variability of plants, animals and microorganisms at the genetic, species and ecosystem level that are necessary to sustain key functions in the agro-ecosystem, and its structure and processes for, and in support of, food production and food security.[2] Agro-biodiversity has spatial, temporal, and scale dimensions, especially at the agro-ecosystem level. These agro-ecosystems are determined by three sets of factors, namely genetic resources, the physical environment and human management practices.

The Ethiopian region is characterised by a wide range of agro-climatic conditions, which account for the enormous resources of agro-biodiversity that exist in the country.[3] The most important of these resources is the immense genetic diversity of the various crop plants.

Doggett suggested that sorghum originated and was domesticated in the north-east quadrant of Africa, most likely in the region of the Ethiopian-Sudan border.[4] The presence of wild sorghums and their cultivated forms, and the ecotype differentiation of sorghum into different races[5] and their presence in different parts of the country, supports the proposition that Ethiopia is the primary centre of origin and hence the centre of diversity for sorghum.

The Ethiopian Institute of Biodiversity Conservation and Research holds about 60,000 accessions of some 104 species of sorghum.[6] Sorghum collections have been made at various times by different organizations in the country and collected close to 8000 accessions.[7] Of these collections, a considerable number are from eastern Ethiopia.

These accessions represent a wide array of diversity and the major sorghum areas in the country. However, the collection is not exhaustive and does not cover the whole country, and a systematic collection and updating is lacking.

It was Harlan and Martini in 1936, as cited by Brush,[8] who recognised the problem of genetic erosion (GE) for the first time. A recent notion of GE is well discussed by Hammer and Laghetti.[9] GE implies that the normal addition and disappearance of genetic variability in a population is altered so that the net change in diversity is negative.[10] Conventionally, GE is the loss of genetic diversity, including the loss of individual genes and the loss of gene complexes such as those manifested in locally adapted landraces.[11] According to Maxted and Hawkes, the loss of genetic diversity is of primary importance for utilization of a crop plant, because mankind relies upon the use of genetic rather than its taxonomic boundaries.[12] In addition, the genetic diversity of a given species expands further than its taxonomic boundaries. GE is detrimental to the short-term viability of individuals and populations, the evolutionary potential of populations and species, and the direct use of genetic resources.[13]

The FAO global plan of action recommends monitoring and developing an early warning system for the loss of plant genetic resources for food and agriculture (PGRFA).[14] The on-farm monitoring of GE in sorghum in Ethiopia is deemed necessary for five major reasons: (i) over the years, there have been changes in cropping systems to adjust to the prevalent complex, fragile, risky and diverse sorghum production environments; (ii) changes in the socio-economic environment have an implication on cropping conditions and thereby on the on-farm management of genetic resources; (iii) the change of climatic conditions in the sorghum-growing environments might have changed the type of varieties used; (iv) the impact of biotic stresses needs to be understood; and (v) formal sorghum research in Ethiopia was started in 1960s with the establishment of Haramaya University. Since then, many varieties have been released for production which might result in the replacement of farmers' varieties. After four decades, it had become imperative to establish whether farmers are maintaining the *de facto* on-farm genetic diversity or whether they have replaced them by IVs.

In view of the above, the objectives of this study were to assess: (i) farmers' perception of GE; (ii) the reasons and levels of GE; (iii) the varieties and areas affected by GE; and (iv) the modalities shaping the process of on-farm GE.

2 MATERIALS AND METHODS

Eastern Ethiopia (Figure 1) was selected for this study for the following reasons: (i) sorghum is the first food crop in the region in terms of area planted, production and importance; (ii) the region is one of the micro-centres of diversity for sorghum, and hence provides ideal sites for studying on-farm GE; (iii) the production of sorghum in diverse ecologies (altitude, rainfall, soil type, landscape, etc.) helps with the assessment of the impact of environmental factors on GE; (iv) the diverse social, cultural and economic conditions prevalent in the region help to tap the Indigenous Technical Knowledge (ITK) associated with the crop; and (v) there are diverse cropping systems, namely, mono-cropping, intercropping associated with pulses and other cereals, and alley-cropping with different perennial crops, and these need a *de facto* genetic diversity to be successful.

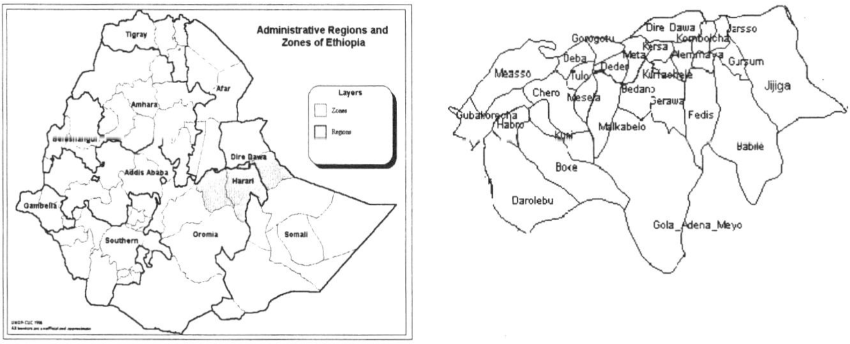

Figure 1 *The position of the study site in Ethiopia and detailed* wereda *map of eastern Ethiopia*

In order to assess farmers' management of on-farm GE, survey research was undertaken. This comprised (i) focused group interviews with 360 farmers (*informal discussion to explore on-farm GE subtopics in depth by interviewing a small, carefully selected group, homogenous in social composition*); (ii) direct on-farm participatory monitoring with 120 farmers (*assessment of genetic resources management with participation of the farmer directly on-farm and in the house*); (iii) key informant interviews (*interviews with special sorghum knowledge holders in farming communities*) with 60 farmers (*farmers seconded by the farming community for their rich indigenous technical knowledge on sorghum production, management and utilization*) and development agents; and (iv) semi-structured interviews with 250 farmers (*survey done with a structured questionnaire with individual farmers for quantifying and comparing data on on-farm sorghum GE*).

Diversity fairs (*on-farm exhibition, description and characterisation of sorghum diversity on a panicle or whole plant or seed*) were also employed for assessing and inventorying on-farm GE. This was carried out around the time of the physiological maturity of the crop. An average of 50 farmers participated in each of the 24 diversity fairs, i.e., a total of 1200 farmers were involved. Both women and men brought all the varieties grown in their field to the fair and discussed prevalence, distribution and the importance of each variety.

Participatory collection of germplasm was done in the year 2000, for comparative estimation of GE with that of 1960. Geographical information of the 1960 collection sites was used to synchronise with the collections made in the year 2000. Both focused group and key informant interviews were made and the folk species and varieties grown in the region were listed.[15] Most of them were collected in the course of the diversity fairs held in the various farming communities across the three sorghum growing ecologies. Besides, additional collections were made in the course of on-farm monitoring for quantification of the number of varieties and to assess the eco-geographical factors affecting GE. In all the directly monitored farms, a participatory zigzag sampling in the diagonal direction of the plot was made with 120 directly on farm-monitored farmers. All the varieties encountered were counted. For varieties in the field that were not encountered in the course of monitoring, the farmers were consulted.

The age of the participating farmers ranged from 18 to 80, with a mean age of 42 years and a farming experience of 20 years. As aforementioned, more than 2000 farmers participated in the study, of which 1400 were male and 600 female. Male-headed one-wife household type (92%) was the dominant one in the region. The participating farmers ethnically comprised *Oromo*, *Amhara* and *Somali* in decreasing order.

In order to quantify GE, Genetic integrity (GI) and GE were calculated as indicated by Hammer *et al.*[16] GI is defined as the ratio of the number of collected accessions per crop per area where farmer varieties (FVs) were presented in 1960 and 2000, and is expressed as a percentage, i.e., GI = (C_{2000}/C_{1960}) x 100%. Information about the 1960s collections was obtained from the Ethiopian Institute for Biodiversity and Conservation. GE was then calculated as GE = 100% - GI. Using the indicated formula, the number of collections made in the 1960s was compared with those collected in this study, in the year 2000. As well as comparing the number of collections made, the names were also compared, and this was appropriate because of the similarity in the area of collection. Comparison by name is used for the first time in this study. Besides the comparison of collections, survey methods such as on-farm monitoring, semi-structured interviews and focused group discussions were also used to assess on-farm GE.

In order to assess the long term trend on sorghum and maize area and production nationally, 42 years crop acreage data was used. The data was obtained from Central Statistical Authority of Ethiopia and the FAO agricultural statistical data base.

Collected data were subjected to descriptive statistics, analysis of variance, log-linear regression, and trend analysis using STATISTICA, SPSS ver. 10 and MINITAB Ver. 14 statistical software.

3 RESULTS AND DISCUSSION

3.1 Farmers' perception of GE

Farmers perceived GE as varietal loss. It is the loss of a variety on the farm (a reduced role or benefit of the variety) owing to biological, economic, ecological and socio-cultural reasons. Sometimes farmers do not grow a variety when it is 'tired off', *i.e.*, not performing very well in terms of yield and resistance against various biotic and abiotic stresses. This does not equate to the conventional sense of GE. Sometimes, it might not mean that the net change (alteration of varietal addition and disappearance) in diversity is negative, as indicated by Brush.[10] There can still be a loss even if the net change in diversity is positive. The number of varieties can be the same but the type of varieties can change, i.e., there can be loss in type but not in number (reduced population size).

GE can happen at the level of single alleles, allele complexes, FVs, whole populations, ecotypes or even at the species level. However, in view of the established culture of growing varietal mixtures and the moderate out crossing rate, it is difficult to presuppose that GE has happened at single allele or allelic complexes. In eastern Ethiopia, sorghum is not as high in the state of genetic vulnerability at a gene level because farmers usually practice varietal mixtures per season. Reduction of GE at the level of gene loss is also enhanced by the level of sorghum out crossing rate of up to 50%

(with a mean of 5%, depending on the various factors).[17-19]

In this study, GE is shown at the variety level. As mentioned above, GE is revealed by the low proportions in the varietal composition and thereby a reduced acreage allocation to the same. However, it was understood that what is reduced or lost at the individual level was not happening at the community, *wereda* or regional level. Even farmers who have stopped growing a particular variety will tell you where or with whom you can find that variety in the same community. In this region it is really difficult to say a variety is lost completely. In the course of on-farm monitoring, it was found that a farmer will tell you that he has lost a variety, but when you go to the field, you will find it there. When you ask the farmer why he had told you he had lost that variety, he will say that yes, this is just '*tokuuma*' (in the Oromiffa language, one of the Ethiopian languages), meaning *'very few plants'*. So farmers attach a reduction in the population of a variety to the loss of the variety, and it might not refer to the complete absence of the variety in their fields. Hence, it is difficult to say that there is a varietal loss in an absolute sense in the centre of diversity. But we can simply concede to what farmers define as varietal loss or reduced varietal importance.

It has been found that for the growth of sorghum in Ethiopia, GE is not a simple, automatic and discrete activity of the farmers in the management of a varietal portfolio. This view is also shared by others, and there is no support for a 'simple replacement theory' in the centre of genetic diversity.[20,21]

3.2 Major reasons for GE

The main cause of GE in crops as reported from almost all countries is the replacement of farmer varieties (FVs) by improved varieties (IVs).[11] But for sorghum in Ethiopia, the dominant cause of GE was not the introduction of IVs. On the contrary, the introduction of IVs has resulted in a widening of the genetic base. The five most important factors for varietal loss of sorghum in the region, in decreasing order of importance, and as indicated in Table 1, were: reduced benefit from the varieties, drought, *khat* expansion, reduced land size and the introduction of other food crops.

Reduced benefits of varieties can be due to low yield, susceptibility to pests and diseases, to *Striga* and so on. Drought has recurrently been affecting the growth of sorghum. In some years, drought becomes serious and wipes out the whole sorghum production, and this might result in the loss of varieties. In other years, drought occurs at various stages, of which the most important is the terminal stress, which results in reduced seed production.[22]

Khat (*Chata edulis* Forsk) is a mild drug crop which originated in Ethiopia. Its growth is expanding at an alarming rate in the region and is taking over the land that used to be used for sorghum. Farmers are aggressively shifting into *khat* production from sorghum and coffee (*Coffea arabica* L.) because of the higher income it brings. Nationally, *khat* is also one of the most economically important crops for export earnings. In the year 1999/2000, *khat* worth close to 500 million USD was exported to many countries.[23] Much of the *khat* production and export is from eastern Ethiopia, the region under study. The estimation for the year 2005 is less than that of 2000. The value for domestic market is expected to be higher than that of the export one.

The effect that a reduction in the area used to grow sorghum, due to the introduction of other crops, has had on GE is discussed in detail in parts 3.4.1 and 3.4.2 of this paper.

Besides, regardless of wealth groups and ecological regions, there was no variation in the loss of varieties. As shown in Table 2, there was a proportional difference in the various wealth groups with respect to the level of GE as average farmers are those with dynamic cropping systems where there is continuous influx and outflux of varieties. However, if you look within each wealth group across various ecologies, the proportional differences are very small (Rich: 4 vs 3.3 vs 3.3; Average: 21.3 vs 28.7 vs 16.7; and Poor: 10.7 vs 9.3 vs 2.7). Intermediate altitudes and average farming domains have more dynamic biological, ecological, social, and economical conditions which led to proportionally higher GE. By contrast, there was variation among farmers and ecological regions for the absence of GE (Table 2).

Table 1 *Factors affecting GE on farms*

Factors	% (N=250)
Reduced benefit derived from the varieties	51.7
Drought	41.4
Expansion of *khat*	9.7
Reduction in land size	9.2
Introduction of other food crops (maize, horticultural crops)	8.5
Lodging problems	3.7
Destruction of habitat and farmland	3.1
Change in land use pattern	3.0
Better sorghum varieties available	2.5
Reduction in the number of sorghum farmers	1.8
Bird damage	1.8
Change in economic status of the farmers	1.2
Late maturity of the varieties	1.2

Table 2 *Proportion of GE in various wealth groups and ecological regions*

Genetic Erosion	Ecological region	Wealth rank			Total
		Rich	Average	Poor	
Yes (NS)	Highland	4	21.3	10.7	36.0
	Intermediate	3.3	28.7	9.3	41.3
	Lowland	3.3	16.7	2.7	22.7
	TOTAL	10.7	66.7	22.7	100
No*	Highland	1.0	35.4	11.5	47.9
	Intermediate	4.2	27.1	5.2	36.5
	Lowland	1.0	14.6	0	15.6
	TOTAL	6.3	77.1	16.7	100

*Significant difference (at 5%) in the proportion of farmers experiencing no GE by wealth and ecology. NS=Non-significant difference. Phi and Cramer's v values were not significant.

Commonly, the advancing pressure on land and labor increases the importance of 'yield' as a selection criterion for farmers, leading to more intensified crop production, the adoption of IVs with higher yield potentials, the use of more inputs and the planting of fewer varieties on a farm.[24] This was not observed to any significant extent in the study area. Besides, in contrast to the idea of de Boef *et al.*,[25] in eastern Ethiopia, the vast reservoir of local knowledge about varieties and their maintenance is not threatened, and nor is the genetic diversity of plants, through the widespread adoption of modern varieties. Hence, farmer varieties did not tend to be left only to the more marginal, risk-prone habitats and ethnological niches for which the IVs were not suitable.

3.3 Quantifying GE

GE is a protracted process in areas where farmers are growing many numbers and types of varieties. Actually, it is easier to assess the loss of a variety in modern agriculture based on IVs in countries which are not at the centre of diversity than in a country such as Ethiopia at the centre of diversity. Besides, GE demands time series data, and it is very difficult to quantify the same in the centre of diversity.

On top of the complexity of the process, the concept of GE is constrained by the cut off value. For instance, what proportions of farmers have to grow a certain variety in order to say that the variety has been genetically eroded? What should be the minimum area on which a variety has to be planted in order to say that the variety is at the brink of loss? Similarly, the most common means of assessing GE in farm level crop diversity is by counting named varieties, but this is different from GE itself because variety names might not correspond to genetic content, either geographically or over time.[24]

Notwithstanding the complexity of assessing GE, I have indicated below the assessment of varietal loss (GE) by various methods. These are: (i) by using a semi-structured interview at individual, community or *wereda* level; (ii) by temporal comparison of 1960 and 2000 collections by number and name; and (iii) by using the area (spatial) method to quantify GE from various perspectives. Finally, the impact of sorghum IVs and those of other food crops on the GE of sorghum is also discussed. It is hoped that this gives the first ever comprehensive and multi-faceted analysis of the GE of sorghum.

3.3.1 GE at individual farmer level

Farmers in general stated that they have not lost all their varieties. If a particular variety is not on their farm, they indicated that it will be found on other farms in the village. They also believe that the area they allocate to sorghum has been tremendously reduced due to a change in the farming system, which has resulted in farmers producing more horticultural crops, food legumes, *khat* and maize. Actually, this pressed them to focus on some varieties of sorghum, rather than trying to grow them all.

Farmers prioritise varieties and reduce the number of varieties grown to respond to a reduction in land size. Some varieties, like *Fendisha, Muyra* and *Wegere,* are avoided by some of the farmers for certain reasons. This has resulted in the ecotypic differentiation[15] of varieties and also in growing specific varieties per farmers' village. However, varieties lost to one farmer will be gained by another farmer in the same or a different community, or a variety lost to a farmer might be found either within that Farmers' Association (FA) or *wereda*. Hence, it is important to determine the level at which GE is happening. As indicated in Table 3, both loss and maintenance occurred in various *weredas* and FAs.

The decision for GE was not collective; it rested upon the individual farmer. The change of varietal portfolio is a continuous and dynamic process at the individual farmer's level. Hence, the varieties lost by an individual farmer will still be grown by the other farmer. In part, this is an indicator of the heterogeneity in socio-economic and bio-physical environments in farming systems in the centre of diversity. This is the most common scenario in the centre of crop diversity and has been observed in cassava,[26] potato,[22,27] maize[28] and rice.[20] The criteria used by farmers to select which varieties of sorghum they should grow varied. On the contrary, for commercial uniform growing of varieties in modern crop production, varietal loss can happen over the region, due to disease epidemics, for instance. This happens commonly with varieties of bread wheat in Ethiopia.

The loss of sorghum varieties across *weredas* and across the whole region is indicated in Table 3. The year 1984 was selected for three major reasons, namely: (i) Ethiopia was hit by severe drought, and this was one of the prime factors for GE; (ii) there was restructuring in rural areas, in particular, villagisation and formation of farmers' co-operatives of socialist orientation system; and (iii) formal sorghum research began in 1960, and the duration, four decades, is enough to assess the impact of modern varieties on GE.

Even if the varieties indicated were lost by a certain proportion of farmers, they are still grown by many others (Table 3). The reason why the answer 'No' was given, except for the variety *Tomis* in *Hirna wereda*, was due to the fact that the varieties indicated as being lost by a certain proportion of individual farmers were still grown by many farmers at the FA and *wereda* level. In addition, the focused group discussions held with farmers at various sites indicated that there was no significant variation in the types of sorghum varieties they grew before 1960, during the period 1970-84, and after 1985, except that they had to focus on some of the varieties and adjust the varietal portfolios according to the prevalent growing environments.

3.3.2 GE by temporal method (time method): GE estimation by comparative quantitative method for the collections of 1960 and 2000

What has been described earlier is the semi-structured survey method for the assessment of GE at the individual farmers' level. The other method that can be used to assess GE is the comparative quantitative method, which is the comparison of amount of collections of two milestone dates. Genetic Integrity (GI) and GE were calculated as suggested by Hammer *et al.*[9] Quantifying the same using the names of varieties is the method suggested for the first time in this study. This was possible as the 1960 collections passport data has included local varietal names.

GI = ratio of the number of collected accessions per crop per site where landraces were presented in 1960 and 2000, = $(C_{2000}/C_{1960}) \times 100\%$. Accordingly, GI by number and name were 167% and 238% respectively (Tables 4 and 5). GE was then calculated as GE = 100%-GI. GE by number and name were –67% and –138% respectively, and

Table 3 *Varieties lost and the reasons for their loss after 1984 (for the period of 1984-2000) at the individual (across the region), FA and* wereda *level*

Varieties	Reason for loss after 1984 at individual farmers level		Overall (%) (N=250)	FA* level	*Wereda*** level
Fendihsa	• Late maturity • Lodging • Pest and disease • Reduction in land size	• Drought susceptibility • Expansion of maize • Decrease in soil fertility • Expansion of *khat*	18.4	No	No
Muyra	• Pest attack • Lodging problem • Late maturity • Drought	• Other better FVs came • Very much consumed as *Eshet* (green), thus lost (not enough seed) • Maize expansion • Not easily threshed	20.4	No	No
Daslee	• Decrease in yield performance • Early maturity and bird damage • Reduced land size	• Drought and pest • Stem breakage at the node	2.8	No	No
Gebabe	• Late maturity • Many nodal tillers	• Drought susceptible • Maize expansion	6.4	No	No
Bullo	• Panicle attacked by pests • Maize expansion through extension	• Pest susceptible • Drought susceptible	6.0	No	No
Wegere	• Pest and disease susceptible • Weevil attacks when allowed to hang over the roof	• Maize extension package expansion • Weevil susceptibility	8.4	No	No
Abdelota	• Other better sorghum FVs came • Pest susceptible	• It is sweet and is eaten by the farmers for its sweet stalk before it matures and thus reduced harvest and less seed	4.4	No	No

Worabi	• Drought • Decline in yield performance	• Other better sorghum FVs came	2.4	No	No
Tomis	• Pest • Yield reduction	• Less storability	3.2	No	Yes-only in *Hirna Wereda*

Wereda** (an administrative structure within the regional state comprising Farmers' Association); Farmers Association* (FA) is lowest administrative structure delimited based on geographical settlement pattern and locations and it is the main power centre to run various activities at the grass root level.

hence, there was no loss in either case. This agrees with the findings of Hernandez, who disproved the GE of maize in Mexico in the centre of diversity.[28] However, when we use the new way of assessing GI and GE, i.e., on the basis of name, twelve varieties which used to be grown in the 1960s were not grown now, even though the number of varietal names in 2000 was greater than that of 1960. In fact, there were many new varieties added in 2000 as compared to that of 1960.

Table 4 *GI and GE based on the number of FVs*

Parameter	Amount
Total number collected in 1960	379
Total number collected in 2000	635
Genetic integrity	167
Genetic erosion	-67%

Table 5 *GI and GE based on the name of the FVs*

Parameter	Amount
Total names*collected in 1960	36
Total names** collected in 2000	86
Genetic integrity	238
Genetic erosion	-138%

*8 names from the 1960 collections were not clear but are added to the total count.
**More than 86 names are identified but only samples of 86 were present.

3.3.3 *Why has there been insignificant GE as quantified by temporal methods?*

1. The farmer-to-farmer seed network in the region has improved the movement of varieties over the years, and it might result in the appearance of some of the diversities that were not captured at the time of the classical collections in the 1960s.
2. Farmers as individuals might have reduced the population in the varietal portfolios. However, at the community, *wereda* and regional level, the net diversity is positive.
3. Farmers as described in the various part of the text are knowledgeable about the importance, management and control of on-farm genetic resources, hence a variety has never been found to be 'useless', though it might have reduced importance.

4. The process of acquiring a new variety is easy, while the reverse process is not the same. There is no simple rejection of a variety as such.

In view of the above, there is no significant GE, either by number or by the names of folk varieties and species. One thing to be noted here is that, whatever the discrepancy in the collections, almost all of the collection in number and name has been maintained on-farm by the farmers for the last four decades or so. These varieties are still under production, though there can be some changes in the varietal performances over the years. The maintenance of these varieties in production indicates the importance farmers attach to their FVs.

In conclusion, maintenance of the varieties for 40 years is a reality and talk about GE is rhetoric.

3.4 GE by spatial method (area method)

3.4.1 Comparison of sorghum production area of 1974 and 2000: Individual farmers level

Sorghum has been grown in Ethiopia for hundreds of years.[29] As time has passed, there have been changes in agricultural systems, for reasons of climate, biotic and abiotic stresses, socio-economic demands and varietal performance. Accordingly, individual farmers adjusted the allocation of the farm size depending on the prevalent circumstances. This was shown by on-farm monitoring of 250 farmers who have been growing sorghum for the last three decades (Figure 2). The net sorghum area has reduced for the period of 26 years, between 1974 and 2000. The mean area for sorghum in 1974 and 2000 were 9.72 *timmad* and 6.47 *timmad* respectively (1 *timmad* = 0.125ha). Hence, there has been a 30% reduction in the area on which sorghum is grown. Part of the land that used to be for sorghum is now used for various other crops, namely potato, horticultural crops, legumes, wheat, barley, coffee and *khat*.

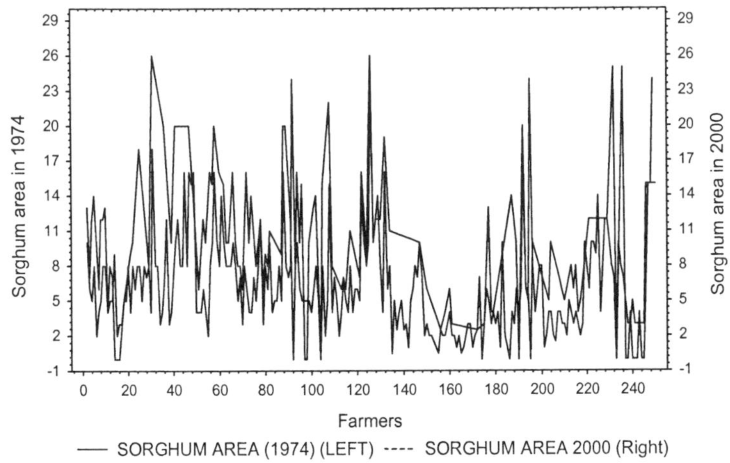

Figure 2 *Comparison of areas allocated to sorghum for the years 1974 and 2000*

3.4.2 Impact of other food crops on GE of sorghum

Both sorghum and maize (*Zea mays* L.) are C_4 plants which compete for similar ecological niches. Throughout the world, sorghum is commonly grown in dry areas. Despite this global scenario, in Ethiopia, sorghum grows in various climatic regions, from humid and dry lowlands to cool and high rainfall areas.

It is believed that maize, the common bean (*Phaseolus vulgaris* L.) and the potato (*Solanum tuberosum* L.) were introduced to Ethiopia in the 16th century by the Portuguese. Emperor Menelik-II introduced maize to eastern Ethiopia at the beginning of the 20th century, so maize has been grown in eastern Ethiopia for about 100 years. Since its introduction, the growth of maize has been expanding in both area and yield, and maize is the second most important crop in Ethiopia in terms of the area grown. Besides, by virtue of the physiological similarity of the two crops, maize has been competing for the areas that were growing sorghum. If we look at the trend for the whole country (Figure 3), the area on which maize is grown has increased by 21.21%, while the sorghum area declined by 2.23%. Maize has taken most of the area that used to be for that of sorghum and for other indigenous crops.

Twelve *weredas* represent the major maize production areas in eastern Ethiopia (Figure 4). As indicated in Figure 5, in these *weredas* maize has taken the place of sorghum production over the years. A continuous expansion of maize in these areas is due to the longer period of sorghum growing, problems associated with bird damage in sorghum, less cold tolerance of some sorghum varieties compared to maize, and the higher yield of maize.

On one hand, maize is replacing sorghum here in Ethiopia; and, on the other hand, sorghum is replacing maize in Mexico, the reverse scenario. The rise of sorghum in Mexico has been so spectacular that it has been called '*the country's second Green Revolution*' and, in fact, Mexico has become the sixth largest sorghum producing country in the world. The drought resistance of sorghum compared with that of maize and wheat has been the single most important reason for the expansion in the growth of sorghum in Mexico.[30]

Turning to other cereals, small-grain cereals such as *tef* (*Eragrostis tef* (Zucc.) Trotter) and bread wheat (*Triticum aestivum* L.) are partly taking over the area of sorghum growing, though not at an alarming rate. *Tef* and wheat were mono-cropped except in some areas where they were intercropped and under-sown with sorghum. For example, in Eastern Hararghe, *tef* was expanding in *Gursum* and *Bedeno*, and wheat in *Kersa, Gorogutu, Melkabello, Bedeno* and *Gerawa weredas*. This expansion in the growth of small-grain cereals is due to a higher demand in the market and the shorter growing duration (double cropping possible), together with *Striga* and bird problems on sorghum. Farmers in the *Hirna* area are increasingly shifting from sorghum to *tef* and other small-grain cereals for reasons associated with drought, length of time to maturity and market requirements. Compared with *khat*, food legumes and small-grain cereals were not threatening sorghum as such, were a source of crop diversification, and could also be grown as bonus crops alongside sorghum.[24,31]

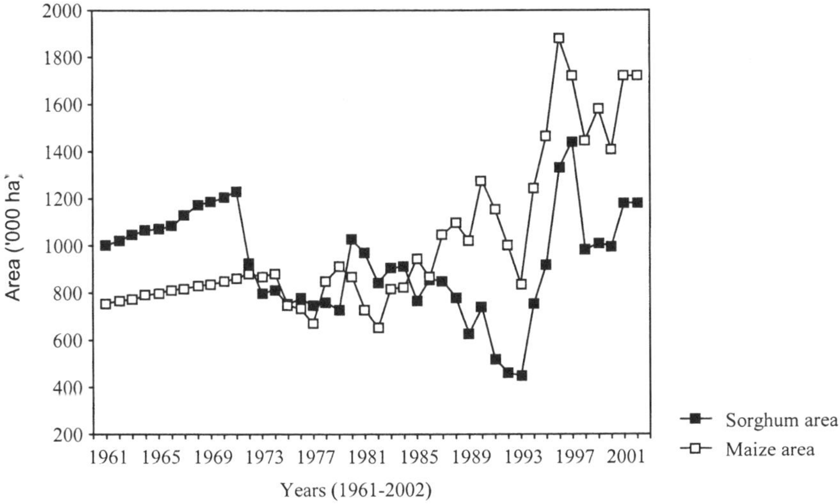

Figure 3 *Long term trend of sorghum and maize production area in Ethiopia. NB: the area allocated for improved varieties of sorghum was only 6,312 ha out of 1,253,620 ha, which is less than* 1%.[32-34]

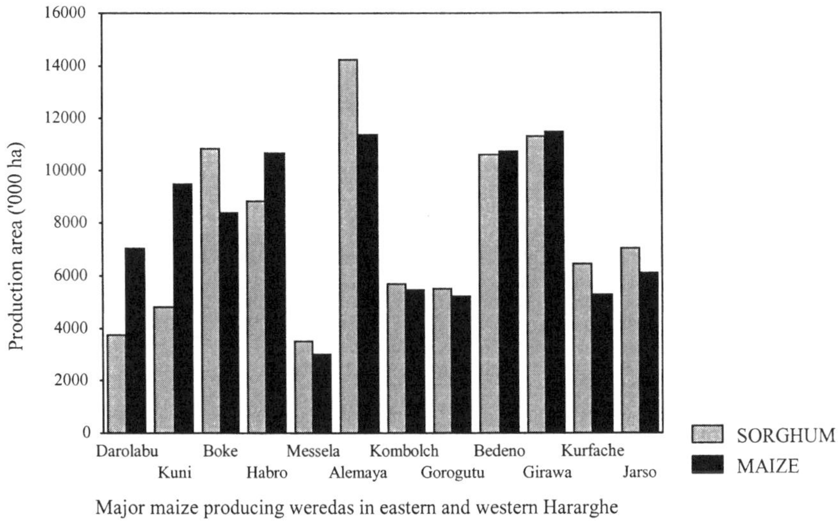

Figure 4 *Area of maize and sorghum production in eastern and western Hararghe weredas identified for sorghum production threatened by maize.* Data source: Bureau of Agriculture for western and eastern Hararghe.

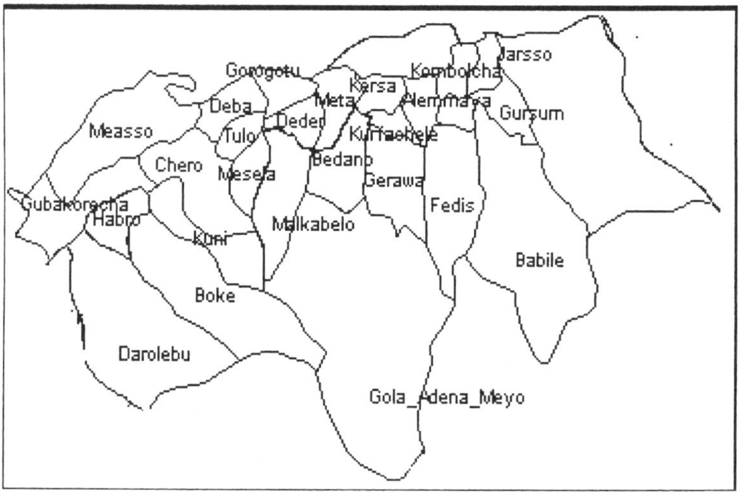

Figure 5 *Shaded* weredas *indicate sorghum production threatened by maize. In particular the cold-tolerant sorghums are threatened by maize in* Gerawa, Kurfachelle, Bedeno, Gorogutu, Messela, Alemaya, Kombolcha *and* Jarso. *The* Bedeno, Girawa, Jarso, Melkabello, Messela *and* Deder weredas *are where high altitude sorghum production has been partly replaced by maize*

3.5 GE as affected by modern sorghum varieties: no infra-specific competition but complementation exists between FVs and IVs in eastern Ethiopia

In eastern Ethiopia, IVs of sorghum are important only in the lowlands, and the number of farmers who have adopted IVs in the intermediate and highland areas is almost nil. Hence, the data below refer to the lowland farmers.

Of 250 farmers interviewed with the semi-structured interview (SSI) questionnaire, and 120 farmers in the on-farm monitoring, only 12% have grown IVs.[24] Adoption of IVs is inversely proportional to the amount of diversity. This is witnessed in eastern Ethiopia in the lowlands, which had comparatively lower diversity vis-à-vis highland and intermediate altitude areas. It has been indicated that Ethiopia has a narrow genetic base for lowland sorghum, and this is one of the many reasons for the adoption of IVs by farmers in the lowlands.[25] Farmers' justification for adopting both FVs and IVs are indicated.[25] As shown previously, none of the farmers who have adopted IVs, according to both the SSI and on-farm monitoring, have completely replaced the FVs. The IVs are increasingly becoming the components of the varietal mixture. Replacement of FVs by IVs could not happen as the reasons for the adoption of IVs and FVs were different. Hence, in the context of eastern Ethiopia, IVs of sorghum are not the reasons for GE of FVs. On the contrary, they have made some contributions to the genetic enrichment of the existing on-farm genetic diversity.

In general, value attributes for yield, seed quality, taste, biomass, feed, and wood for fuel and construction are ranked higher for FVs, while for early maturity and drought resistance IVs were ranked higher.[25] There were also some FVs that appear to be associated with better drought resistance attributes.

On the contrary, improved crop and pest management measures don't lead to the cultivation of IVs in the region. Actually, sorghum farming in eastern Ethiopia is a natural crop farming in which farm labour is the input most of the time. The current sorghum extension package program in the region used FVs and other improved farm management practices. The FVs have been ratified by the regional bureau of agriculture as being better than the available IVs. The most amazing scenario of adoption is that there is no simple replacement of FVs by IVs. The most common process is addition, not substitution.

The reasons for farmers maintaining their varieties while adopting improved ones were as follows:

1. Farmers knew that the IVs did not meet all their requirements.[25]
2. Poor seed quality and a lower multipurpose value of IVs forced farmers to retain their own varieties.[25,37]
3. The IVs are normally used as 'emergence varieties' in the event of failure of the first (normal) planting. Hence, early and drought-resistant IVs are planted to capture the crop season.[24]
4. Analysis of the yields of IVs and FVs cultivated in this region showed that FVs were actually quite competitive and higher yielding than IVs.[24]
5. FVs have better pest and disease resistance than IVs, and this was one of the factors for the maintenance of FVs while adopting IVs.[20]

In summary, the prediction in the late seventies that there would be complete erosion of FVs by IVs by the end of the eighties;[35] and the principle of GE of Frankel[36] and Harlan[37] that competition between IVs and FVs favours the former and results in the replacement of the latter, are not valid in the context of sorghum in eastern Ethiopia. In view of these facts, there was not any infra-specific competition between FVs and IVs, rather there is infra-specific complementation between IVs and FVs of sorghum.

3.6 The Modalities for GE

The concept of GE appears to be simple but this is deceptive. The process of conservation and replacement is normally an outcome of the interactions between socio-economic and biophysical environmental factors and the action of both natural and conscious selection.[8] To explain the process shaping the GE scenario on-farm, three modalities have been suggested. Actually, Brush indicated that GE of crop populations is an ecological concept that draws implicitly on models of ecology and population biology,[10] but this concept has not been theoretically elaborated or empirically tested for crops. Hence, attempts are made here to use ecological and farmer management models, and their interaction, to explain the GE scenario in eastern Ethiopia. The models indicated below are the function of both human and ecological concepts. However, the ecological, Meta-population and Niche models *per se*, as suggested previously,[6] cannot comprehensively explain the whole scenario of GE on-farm.

Hence the following three new models are suggested to explain the scenario:

1. Bioecogeographic induced GE model: this model embodies the natural factors affecting GE, namely, biological, ecological and geographical factors. Drought, bird damage, late maturity, pests and disease, and *Striga* are some of the bioecogeographic factors that farmers indicated as being responsible for GE on sorghum in the region (Tables 1 and 3).

This model generally takes into account the natural ecological model. The niche model proposed by Brush refers much to that of interspecies interaction that is only a part of this model and it explains only the crop introduction and enhancement aspects.[8] Intra-specific competition is not indicated by the model, though it was suggested to be scaled down for inter-varietal competition. Inter-varietal competition that leads to a balance or imbalance in the specific niches is niche theory for GE but this is not sufficient. The assumptions of the classical models are easily violated in farming systems in the centre of diversity, so that we should expect co-existence rather than the competitive exclusion to prevail. Hence, the niche model does not completely explain the GE scenario.

The other model, namely the meta-population model, is the island biogeography concept, which refers to reduction of area and fragmentation of populations, threaten species' survival and biological diversity.[8] This also explains only that of the land size as a factor for GE model. Both the niche and meta-population theory are the part of bioecogeographic enhanced GE model. In the meta-population model, particular populations may not be so important as long as their alleles and agro-morphological characteristics are present in other linked populations. This useful notion is included in the model, but it is not easy to verify.

In general, both the models did not include the human aspect and the interaction of human and natural aspect, and the models suggested here better explain the on-farm GE scenario.

2. Farmer induced GE model: this model refers to the human factors (Tables 1 and 3) involved in varietal selection, such as changes in food habits and preferences, in the market value of varieties, in production objectives (feed, food, fuel-wood, etc.,), and other socio-economic factors. On top of this, changes in the size of a farmer's household dictates the mixture of crops and livestock that he grows which, in turn, affects the maintenance of genetic diversity. In addition, the presence of various ethnic groups in the region, such as *Oromo, Amhara, Somali, Harare* and *Argoba*, warranted the selection of certain varieties suited to the specific ethnic needs. The range of variations on the type and number of varieties across individual farmers and communities is partly explained by the variation among socio-economic and cultural variations for GE. This is the human-induced aspect of the GE model. Hence, GE is not driven completely by either demand or supply. In some cases, it is the former, in others it is the latter, and in still others, it is both.

3. ***Farmer-cum-bioecogeographic induced GE model:*** This is a blend of human and natural factors. Factors such as reduced benefits from the varieties, the introduction and expansion of other crops (for example *khat* and maize), a reduction in the number of farmers, and the destruction of habitat and farmlands all include both natural and human factors (Tables 1 and 3). For example, reduced benefits, which might be due to low yield, can arise from pest or disease infestation or a poor selection of varieties. Similarly, the

expansion of *khat* in the region is due to adaptation of the crop and farmers' interest in the increased income that comes from *khat*. Similarly, the reasons for maize expansion are its biological competitiveness and the selection of maize by the farmers for a variety of benefits. Hence, expansion and introduction of other crops can be due to natural or human factors. The relative contribution of both factors for GE might vary based on the actual scenario of socio-economic and crop-growing environments.

4 SUMMARY AND CONCLUSIONS

GE is one of the most fascinating issues in the study of the management of plant genetic resources. However, it is not easy to ratify. In this study, the very first attempt has been made to use various comprehensive methods to assess on-farm GE. The area (spatial) method partly indicated the possible presence of GE, but this method is constrained by the lack of evidence for its direct association with GE. However, in the other methods there can be GE at individual farmer level, but at a regional level the net GE was almost nil.

In order to corroborate the area method for GE, comparative evaluation of varieties conserved *ex situ* and *in situ* (on-farm) needs to be made at the molecular and isozyme level. The 1960 and 2000 collections contain many similar varieties from the same sites. Farmers have witnessed that there has been a change of varietal performance but not significant loss of varieties over years. Hence, the farmers' assertion has to be supported by empirical evidence.

The region is endowed with various valuable varieties for increased popularisation and utilization thereby to reduce GE. This needs to be considered by follow up projects in order to support the farmers in the region. Biodiversity is not an end in itself, but is the way to ensure food security, environment and health and enhancement is one way to empower farmers' achievements, increased utilization of on-farm genetic diversity and reduce GE.

The region is partly characterised by seed insecurity. This can be ameliorated by establishing community seed banks at the FA level, and these can also be used for the minimization of GE and improving on-farm genetic resource management and utilization.

Very limited studies have been carried out on the wild progenitors of sorghum. Most reports are limited to indications of the presence of the progenitors. Hence, comparative introgression study of the wild, cultivated and their crosses is imperative as a part of the genetic resource management of the genus.

The complementary role played by both farmer and formal sorghum genetic resource conservation for sustainable utilization is tremendous. In order to benefit from their complementarities, integrated genetic resource management, and utilization for reducing GE has to be implemented.

Acknowledgements

I thank the farmers of Ethiopia for their participation and for sharing their ideas and knowledge on the different aspects of this GE study. I am grateful for the research and field assistants that assisted in interviewing the farmers and in field and laboratory work.

Thanks go also to various NGOs and GOs who have assisted in the various aspects of the fieldwork. This research was supported by the Consultative Group for International Agricultural Research (CGIAR) Participatory Research and Gender Analysis (PRGA) Small Grant Program, the Norwegian government, and Alemaya University. Due thanks go to all for funding the project. The support provided by Haramaya University for the research work in Ethiopia is appreciated. I am grateful to many colleagues who have read and suggested invaluable comments on the first draft of this manuscript. Finally, thanks go to the anonymous reviewer for professional comments and suggestions.

References

1 UNEP (United Nations Environmental Program). 1992. Rio Declaration, World Conference on Environment and Development, United Nations Environmental Program, Brazil
2 FAO. 1999. Sustaining agricultural biodiversity and agro-ecological function. Rome: Food and Agriculture organisation of the United Nation. http::www.fao.org/WAICNET/FAOINFO/SUSTDEV/Epdirect/Epre0063.htm).
3 M. Worede 1992. Ethiopia: a gene bank working with farmers. Pp. 78-94. in Growing diversity (D. Cooper, R. Vellve and H. Hobbelink, eds.). Intermediate Technology Publications. London.
4 H. Doggett 1988. Sorghum. Longman. 2nd ed. Green Co. Ltd. London. 512 pp
5 F. Mekbib, *Journal of New Seeds*. 2008a, 9,43.
6 IBCR. 2000. Draft Strategy of National Biodiversity Conservation and Research Program of Ethiopia
7 Plant Genetic Resources Centre/Ethiopia. 1986. Ten years of collection, conservation, and utilization, 1976-86. PGRC/E, Addis Ababa, Ethiopia
8 S.B. Brush 1999. GE of crop populations in the centres of diversity: A revision. In: FAO-Proceedings of Technical meeting on the methodology of the FAO World Information and Early Warning System on Plant Genetic Resources, Research Institute of Crop Production, Prague, Czech Republic, June 21-23, 1999
9 K. Hammer and G. Laghetti, *Gen. Resour. and Crop Evol.*, 2005, **52**, 629.
10 S.B. Brush 2000. The issues of *in situ* conservation of Crop Genetic Resources. Pages 3-28. In (Ed. S. Brush: Genes in the field: on farm conservation of crop diversity. IPGRI and IDRC, Lewis Publishers).
11 FAO. 1998. The state of the world's plant genetic resources for food and agriculture.
12 N. Maxted and J.G. Hawkes, 1997. Selection of target taxa. In: Plant genetic conservation: the *in situ* approach (ed. Maxted, N., Ford-Lloyd, B.V. & Hawkes, J.G.). pp. 58-98. Chapman & Hall, London.
13 A. Brown, A. Young, J. Budon, L. Christides, G. Clarke, D. Coates and W. Sherwin, 1997. Genetic indicators for state of the environment reporting. Australia: state of the environmental technical paper series (environmental indicators), Department of Environment, Sport and Territories, Canberra.
14 FAO. 1996. Global plan of action for the conservation and sustainable utilization of plant genetic resources for food and agriculture and the Leipzig Declaration. International conferences on plant genetic resources. FAO, Rome.
15 F. Mekbib. Journal of Ethnobiology and Ethnomedicine, 2007a, **3**, 1

16 K. Hammer, H. Knüpffer, L. Xhuveli and P. Perrino, *Gen. Resour. and Crop Evol.*, 1996, **43**, 329
17 L.R. House, 1992. A guide to sorghum breeding. 2nd edition. ICRISAT, Patancheru, P.O. Box, Andhra Pardesh, 502 324, India.
18 P.E. Arriola, *African Crop Science Journal*, 1995, **3**, 153.
19 Y. Djè, M. Heucrtz, M. Atei, C. Lefèbvre and X. Vekemans, *Euphytica*, 2004, **138**, 205.
20 M.R. Bellon, J.L. Pham and M.T. Jackson, 1997. Genetic conservation: a role for rice farmers. In: N. Maxted, B.V. Ford-Lloyd and J.G. Hawkes (eds.), Plant Genetic Conservation – the *in situ* Approach. Chapman & Hall, London, pp. 263-289.
21 S.B. Brush, H. Carney and Z. Huaman, *Econ. Bot.*, 1981, **35**, 70.
22 F. Mekbib, *Euphytica*, 2006, **152**, 163.
23 The Reporter. 2000. Vol. V. No.219, Nov. 15, Addis Ababa, Ethiopia.
24 M. Lipton and R. Longhurst, 1989. New seeds and poor people. Baltimore, MD, USA, Johns Hopkins University Press.
25 W. de Boef, K. Amanor, K. Wellard and A. Bebbington, 1993. Cultivating knowledge: genetic diversity, farmer experimentation, and crop research. London: Intermediate Technology Publications.
26 J.B.M. Sambatti, P.S. Martins and A. Ando, *Brazil. Econ. Bot.*, 2001, **55**, 93.
27 K.S. Zimmerer and D.S. Douches, *Econ. Bot.*, 1991, **45**, 176.
28 X.E. Hernandez, 1993. Genetic resources of primitive varieties of Mesoamerica, *Zea* spp., *Pheasolus* spp., *Capsicum* spp., *Cucurbita* spp. In: Survey of crop genetic resources in their centres of diversity, First Report., O.H. Frankel, ed. Rome: Food and Agricultural Organization of the U.N. Pp: 76-86.
29 F. Mekbib, *Journal of Ethnobiology and Ethnomedicine*, 2008c (in press).
30 NAS (National Academy of Sciences, USA). 1996. Sorghum. In: Lost Crops of Africa. Volume I. Grains.
31 F. Mekbib, *Expt. Agri.*, 1997, **33**, 399.
32 F. Mekbib, *Journal of New Seeds*, 2007b, **8**, 63.
33 F. Mekbib, *Journal of New Seeds*, 2008b, **9**, 191
34 Central Statistical Authority (Federal Democratic Republic of Ethiopia). 2005. Agricultural Sample Survey. Volume IV. Report on land utilization.
35 S.B. Brush, *Crop Sci.*, 1995, **35**, 346.
36 O.H. Frankel, 1970. Genetic conservation in perspective. In: Genetic Resources in Plants – their exploration and conservation. O.H. Frankel and E. Bennett, eds., Oxford: International Biological Programme. Handbook No.11. Blackwell Scientific Publications. Pages 469-489.
37 J.R. Harlan, *Science*, 1975, **188**, 618.

DELIVERY OF APPROPRIATE CATTLE GENOTYPES TO EASTERN AFRICAN SMALLHOLDER FARMERS THROUGH IN-VITRO EMBRYO PRODUCTION TECHNOLOGIES - THE TECHNICAL PROCEDURES, PROSPECTS AND CHALLENGES

H.M. Mutembei,[1,2] S.E. McClintock,[1*] B. Muasa,[2] R. Origa,[2] Jimbo,[1] S.V.T. Tsuma,[2] E.R. Mutiga,[2] A.E. McClintock[1*] and A.M. Okeyo[1]
[1]International Livestock Research Institute (ILRI), Nairobi, Kenya
[2]Clinical Studies Department, University of Nairobi, Nairobi, Kenya
[*]Former affiliation

1 INTRODUCTION

The on-going world food crisis and the rise in demand for food of animal origin, competitive global markets for such foods and the negative effects of climatic change, especially in the marginal tropical areas,[1] demands that innovative strategies and interventions be urgently employed to increase productivity.[2] In developing countries, the dairy sector, especially in low input production systems, has the potential of providing cost effective animal protein requirements. To do this, such units need to be efficient, and the starting point is using appropriate genotypes.

Many attempts aimed at improving the productivity of indigenous dairy cattle in developing countries through crossbreeding with temperate breeds have generally not been very successful.[3] Many studies have shown that temperate dairy cattle breed best when combined with local tropical breeds at 50% of each breed composition (i.e. F1 cross). The F1 is well adapted and produces reasonable amounts of milk under the local conditions.[3-5]

Aggressive promotion of a few exotic cattle breeds in Sub Saharan Africa (SSA) has resulted in cattle genotypes that are not only poorly adapted to the production systems but are also commercially inefficient.[6-8]

In order to meet the increasing demand for appropriate improved dairy cattle genotypes in the marginal dairy systems in Eastern African region, more efficient production and delivery of improved germplasm are needed. Breeding strategies that utilise a combination of emerging reproductive technologies, especially the in-vitro embryo production (IVEP) and semen sexing, offer prudent possible options.[9] For example, indigenous breeds such as the Zebu (Boran) or Sanga (Ankole) that are inferior commercial milk producers would be valuable as oocyte donors as well as surrogates for crossbred embryos under IVEP-based systems. This paper presents potential application of IVEP technologies in Eastern Africa region, in terms of the technical procedures and the challenges involved.

2 MATERIALS AND METHODS

2.1 Brief overview

Production of embryos in the laboratory involves four major steps: 1) Ovary collection, 2) Oocyte maturation, 3) in vitro fertilization and 4) Embryo culture. In brief, ovaries are collected in the slaughterhouse from cows whose pre-slaughter parameters are assessed and recorded. In this study, collection was from exotic dairy and local Zebu cows with body condition scores that ranged from 2.5 to 3.5 and certified as being as being in good health.

The target of the project is to produce F1 crosses using sexed Jersey and Ayrshire semen to fertilize Boran oocytes. However, in this initial test phase, non-sexed Boran semen was used to fertilize ova from exotic dairy breeds; and sexed Aryshire or Jersey semen to fertilize Boran ova.

The ovaries were delivered to the laboratory within 4hrs of collection in a PBS or saline media and at 35-37°C temperatures. The ovaries were collected between 06.30-08.30 hrs, placed in a thermos flask containing Dulbecco's phosphate buffered saline that was supplemented with 0.5ml/L of penicillin (250,000 IU) and streptomycin (0.25g) (Dulbecco's PBS+), pre-heated to 39°C.

2.2 Cumulus oocytes complex (COC) aspiration and maturation

In the laboratory, the ovary temperatures were noted on arrival, they were then washed twice using Dulbecco's PBS+ and then the follicular fluid aspirated by puncturing follicles of diameters of 1 to 8mm using a 18G × 1½″ needle (1.20 x 40mm).

The follicular liquid was then collected in sterilized tubes previously pre-warmed to 37°C and left to stand for 10 min in a water bath at 37°C to settle. The oocytes were then picked up from the follicular fluid into the TALP-hepes (tyrodes-albumin-pyruvate-hepes) medium in a Petri-dish using a Gilson pipette fitted with 1000μl tips, then washed twice in TALP-hepes and once in TCM-199 *in vitro* maturation medium (IVM). Cumulus oocyte complexes (COCs) were selected and placed in a sterile Nunc Petri-dish containing 400 μl of IVM media supplemented with 20ug/ml of FSH, LH and Estradiol and 10% foetal calf serum (FCS). Only COCs with compact multilayered cumulus complex and evenly granulated cytoplasm were selected for *in vitro* maturation. IVM incubation was done under high levels of hygiene and maintained at 39°C, 5% CO_2 and 95% humidity for 22-24hrs.

2.3 In-vitro fertilization (IVF)

In-vitro fertilization (IVF) using frozen Jersey semen was carried out as described previously.[10] In brief, certified semen was obtained through a 50-minutes swim up technique using 0.25 ml of thawed semen that was gently released at a 45° angle to the bottom of a 10ml Sterlin test tube containing 1.5 ml of swim up (SPTL) capacitation medium (TALP) supplemented with 6mg/ml of bovine serum albumin (BSA). Approximately 1.3 ml of the supernatant (the SPTL medium with vital sperm) was transferred into a 10ml Sterilin test tube and centrifuged at 3000 RPM for 5min at room temperature, after which the sperm cell pellet located at the bottom of the falcon tube was re-suspended with 100 μl of fertilization (Fert-Talp) medium. The sperm concentrations during fertilization were 1-2 x 10^6 spermatozoa cells/ml.

The matured oocytes were visually judged on the basis of their external characteristics (e.g. expansion of the cumulus cells and presence of smaller and darker oocytes), as indicators of degree of maturity. The selected oocytes were then placed in a culture Petri-dish containing TALP-hepes medium, washed two times in TALP-hepes and one time in Fert-Talp medium. 15-20 oocytes were transferred into the 50μl droplet of Fert-Talp media that was supplemented with caffeine and Heparin. Then a final volume of the prepared semen, containing 1-2 x 10^6 sperm cells/ml, was added to the 50μl droplet containing the matured oocytes. The *in-vitro* fertilization was performed in media supplemented with 20 μg of heparin/mL and 6 mg/mL of fatty free-acid BSA fraction, covered with mineral oil for 18 h in a humidified atmosphere of 5% CO_2 and at 39°C in air.

2.4 In vitro culture (IVC)

After fertilization, the zygotes were transferred into a Petri-dish containing TALP-hepes medium. They were then washed twice in TALP-hepes and once in culture medium that is supplemented with 10% FCS and 6mg/ml BSA (CR2). Then the zygotes were co-cultured with their cumulus for 72 hrs in a total volume of 100 μl in droplets of the semi-defined CR2 culture medium that were overlaid with mineral oil in an incubator at an atmosphere of 5% CO_2, at 39°C and 95% humidity. Thereafter, half of the medium was drawn out and replaced with fresh medium. Cleavage rate was evaluated and documented after 72 hours of incubation and blastocysts were judged on days 7, 8 and 10 post-fertilization.

3 RESULTS AND THE CHALLENGES

3.1 The experiences to-date

The results reported here are for the non sexed semen-in vitro fertilization-embryo transfer aspects in preparation for the next phase, when sexed instead of non-sexed semen will be used. In the sexed semen phase, which is underway, and the results of which are not reported here, sexed semen will be used in an in vitro fertilization (IVF) set-up to produce sets of F1 female embryos of specific genotypes for transfer to pre-primed Boran cows.

The statistics for the laboratory findings presented at the conference are summarized in Table 1. Ovaries from cows with pre-slaughter condition scores below 2 produced fewer aspirable COCs of low quality. On average, 55% of the ovaries collected were from the local zebu cattle and the rest were from exotic dairy breeds. 10-15% of cows slaughtered were pregnant but with comparable patterns of ovarian follicular dynamics to their non-pregnant counterparts. 50% of the aspirated ova were of high quality (grade A) (i.e. brown homogenous cytoplasm and with 2-3 cells of full cumulus); 35% of grade B (brown homogenous cytoplasm and with some degree of cumulus denudation) and the rest (15%) were either immature or of unacceptable quality due to denuded cumulus or dark heterogeneous cytoplasm. The sizes of ovaries from local zebu cows were unique in that they were smaller in size and had 10-15% fewer recoverable oocytes compared to their exotic dairy breed counterparts.

Table 1 *Summary of some of results obtained during the initial laboratory trials*

Cows' body condition score	Temperature of ovaries on arrival	Number of ovaries collected daily	Number of oocytes per ovary	Number of oocytes incubated daily	Maturation rate %	Cleavage rate %	Blastocyst rate %	Hatching rate %
1.5-3.5	27-33°C	5-18	0-70	100-200	65-90	60-65	25-40	25-35

3.2 Challenges

The main challenges have been inconsistent availability of the appropriate cow genotypes at slaughterhouses. The SIFET system relies on a regular supply of ovaries (slaughter cows) of known and desired genetic composition or breed. Not every cow being slaughtered would therefore be an ideal donor. Genetically superior, healthy and medium aged cows are preferred as donors, yet few such cows are slaughtered. Also, it would help in planning of the IVEP process if the numbers of cows of the appropriate/desired genotypes that are due for slaughter on a given day were known forehand or could be predicted with high degree of certainty. Under local conditions, such information is not readily available, thus raising the associated fixed costs. In addition, the physiology of indigenous Boran cows is slightly different from those of the exotic dairy breeds for which IVEP protocols have been standardized. Modifications and tweaking of the protocols are needed, which take time and additional research.

It would be nice if sexed Boran semen were available so that reciprocal F1 (Boran-Jersey and Jersey x Boran) crossbred females could be both produced. Unfortunately, no sexed Boran semen is currently commercially available, at least at reasonable prices. Our attempts to ship Boran semen to Korea at concessional costs, have them sexed and then shipped back, were futile due to restrictive import and export sanitary requirements.

Farmers interviewed as to their views on IVEP systems based on slaughtered cows as oocyte sources have raised ethical and cultural concerns and showed some concerns over producing embryos from dead cows.

4 DISCUSSION

4.1 General discussion

As reviewed before,[11] differences in results obtained in IVEP systems are caused by various factors such as time taken to transport ovaries to lab after collection,[12] temperature changes during ovary collection and transport to the laboratory,[13] cow breed,[14-16] oocyte quality[13] and follicular, maturation, fertilization and embryo culture environments.[17] In particular, cattle breed has been shown to influence developmental competence of oocytes and blastocyst rates.[15,16] The effect of breed on oocyte quality is amplified by environmental conditions. For example, although Holstein cows (*Bos taurus*) have bigger ovaries, they produce oocytes of lower quality compared to Brahman cows (*Bos indicus*).[18] However, for both cases, the *in-vitro-fertilized* embryos at the 4- and 8-cell stage when subjected to heat stress results in decreased blastocyst rate but the effect is more severe in Holstein embryos compared to zebu embyros.[19,20]

Boran cattle are only characterized with regards to luteal response to luteolysing agents during synchronization programs.[21-25] Because the follicular and IVEP reproductive profiles for the Boran cattle are not well documented, it is has been challenging to come up with an optimum IVEP system for Boran ova. Therefore, reproductive profiling of the Boran cattle, especially in the IVEP related processes, is necessary.

While the post-fertilization culture environment is the most critical determinant of blastocyst quality,[26,27] the production of quality oocytes, with high competency and probability of undergoing full cytoplasmic maturation,[28] is essential for achieving high blastocyst yield in an IVEP system. The results obtained so far have clearly demonstrated that IVEP protocols can be optimized for production of transferable embryos of appropriate genotypes that best match local situations.

Given that, globally, IVEP is currently one of the most important biotechnologies in cattle breeding,[10,29] its potential for delivering appropriate cattle genotypes to smallholder farmers in marginal dairying production systems in SSA is great. In low input systems that are characterized by scarce land and feed, low husbandry skills, poor animal health and breeding services and high heat stress, use of crossbreds, especially the 1st filial generation (F1s) resulting from breeding indigenous cattle to exotic pure breeds, offer better opportunities for sustainable commercial production of both milk and beef, especially the former,[4,10] more so if markets for meat and milk are expanding. Well planned IVEP systems provide options for producing F1s more rapidly and more efficiently and should therefore be embraced.[2]

On the other hand, where some productive and adapted genotypes or breeds have already been developed, in and for environments that are similar to local SSA production environments, IVEP technologies can be used to enable quick introduction and multiplication of such breeds to the SSA environments. Example cattle breeds which can be introduced and multiplied in marginal parts of SSA include the Gyr from Brazil and the Kenena and Butana breeds from Sudan to marginal dairying areas in the rest of Eastern Africa. In both circumstances, the IVEP technology would enable continued and sustainable existence (conservation) of local breeds, Boran in this case, and the moved breeds (Butana, Kenana and Gyr), well beyond what would otherwise be the case.

4.2 Prospects

Optimization of the non-sexed-IVEP protocols using oocytes sourced from the indigenous Boran cattle has been successfully undertaken. The plans to use sexed semen for IVF and use of the ovum pick up (OPU) technique to harvest ova from live animals are underway. OPU requires a special ultrasound machine, which is costly (US$ 3,850-6,500) and will have to be purchased. The procedure can be performed repeatedly on the same animal each week and on older animals that would otherwise not produce a calf either naturally or through conventional ET techniques. OPU can still be used as well as pregnant cows and in calves as young as five months old. With appropriate links to the private commercial ranchers and investment by traders in improved genetics, the prospects for a day-old chick supply model is possible, whereby F1 heifers are contractually produced by specialized firms/farms and sold to small-holder farmers in medium to low potential dairy production systems in Eastern African region. OPU also removes the concerns associated with the use of dead cows as sources of embryos.

5 CONCLUSIONS

The results obtained so far clearly demonstrate that in-vitro oocyte maturation (IVM) of embryos obtained from ovaries of slaughtered cows, subsequent in-vitro fertilization (IVF) of these, in-vitro embryo culture (IVC) and embryo transfer process is a possible option for generating specific cattle genotypes for specific production systems in Eastern Africa. However, there is need for fine tuning of the IVM, IVF and IVC protocols for the local zebu breeds. Planned application of combinations of these technologies would deliver the desired genotypes and commercial business approaches are needed to deliver this technology beyond the current laboratory trial. Ovum pick up (OPU), if embraced, would even make the process much more economically and ethically attractive.

Acknowledgements

The authors acknowledge Heifer Project International for funding this project, University of Nairobi for granting staff time and recipient animals for transfer, International Research ILRI for providing administrative support and recipient animals for transfer, Emrapa scientists, in particular Drs. Luiz, Viana and Sa for their technical advice, and Dogoretti Abattoir and donor cow owners for their co-operation.

References

1 P.K. Thornton, P.G. Jones, T. Owiyo, R.L. Kruska, M. Herrero, P. Kristjanson, A. Notenbaert, N. Bekele and A. Omolo, *Report to the Department for International Livestock Research Institute (ILRI),* 2006.
2 C. Seré, A. van de Zijpp, G. Persley and E. Rege, *Animal Genetic Resources, Information Bulletin,* 2008, **42**, 3-27.
3 E.P. Cunningham and O. Syrstad, *Crossbreeding Bos indicus and Bos taurus for Milk Production in the Tropics. FAO Animal Production and Health Paper,* 1987, **68**. Food and Agriculture Organization of the United Nations, Rome.
4 J.E.O. Rege, *Proceedings of the 6th World Congress on Genetics Applied to Livestock Production,* 1998, Armidale, Australia, **25**, 193-200.
5 J.M. King, D.J. Parsons, J.R. Turnpenny, J. Nyangaga, P. Bakari and C.M. Wathes, *Animal Science,* 2006, **82**, 705.
6 O.L. Bondoc, C. Smith and J.P. Gibson, *Anim. Breed Abstracts,* 1989, **57**, 819.
7 A.J. Mohammad, B.M. Swallow and J.E.O. Rege, *Outlook on Agriculture*, 1999, **28**, 239.
8 A.K. Kahi, G. Nitter, W. Thorpe and C.F. Gall, *Liv. Prod. Sci.*, 2000, **63**, 53.
9 S.E. McClintock, R. Ouma, I. Baltenweck, A.M. Okeyo, A.E. McClintock and J.E.O. Rege, *Proceedings of Association for the Advancement of Animal Breeding and Genetics* 2007, Australia.
10 L.S.A. Camargo, J.H.M. Viana, W.F. Sá, A.M. Ferreira, A.A. Ramos, C. Freitas and V.R. Vale Filho, *Reprod. Fert. Dev.,* 2006, **18**, 243.
11 P. Lonergan, T. Fair, D. Corcoran and A.C. Evans, *Theriogenology,* 2006, **65**, 137.
12 J.G. Thompson, *Reprod. Fert. Dev.,* 1997, **9**, 341.
13 P. Holm and H. Callesen, *Reprod. Nutr. Dev.,* 1998, **38**, 579.
14 D. Lechniak, D. Cieslak and J. Sosnowski, *Theriogenology,* 1998, **49**, 779.

15 A.E. Fischer, D.P. Bernal, C. Gutierrez-Robayo and J.J. Rutledge, *Theriogenology*, 2000, **54**, 1433.
16 A. Boediono, T. Suzuki and R.A. Godke, *Anim. Reprod. Sci.*, 2003, **78**, 1.
17 N.K. Khurana and H. Niemann, *Theriogenology*, 2000, **54**, 741.
18 A. Rocha, R.D. Randel, J.R. Broussard, J.M. Lim, R.M. Blair, J.D. Roussel, R.A. Godke and W. Hansel, *Theriogenology*, 1998, **49**, 657.
19 J. Block, C.C. Chase Jr. and P.J. Hansen, *Mol. Reprod. Dev.*, 2002, **63**, 32.
20 F.F. Paula-Lopes, C.C. Chase Jr., Y.M. Al-Katanani, C.E. Krininger 3rd, R.M. Rivera, S. Tekin, A.C. Majewski, O.M. Ocon, T.A. Olson and P.J. Hansen, *Reproduction*, 2003, **125**, 285.
21 E.R. Mutiga, E. Mukasa-Mugerwa and A. Tegegne, *J. of Agricultural Science (Cambridge)*, 1993, **120**, 103.
22 E.R. Mutiga, E. Mukasa-Mugerwa and A. Tegegne, *The Kenya Veterinarian*, 1993, **17**, 5.
23 A. Tegegne, A.C. Warnick, E. Mukasa-Mugerwa and H. Ketema, *Theriogenology*, 1989, **31**, 361.
24 A. Tegegne, K.W. Entwistle and E. Mukasa-Mugerwa, *Theriogenology*, 1992, **37**, 1005.
25 A. Tegegne, R. Franceschini and S. Sovani, *Theriogenology*, 1994, **41**, 1653.
26 D. Rizos, A. Gutierrez-Adan, S. Perez-Garnelo, J. De La Fuente, M.P. Boland and P. Lonergan, *Biol. Reprod.*, 2003, **68**, 236.
27 D. Rizos, F. Ward, M.P. Bol and P. Lonergan, *Theriogenology*, 2001, **56**, 1.
28 M.M. Bevers, S.J. Dieleman, R. van den Hurk and F. Izadyar, *Theriogenology*, 1997, **47**, 13.
29 M. Thibier, *Data Retrieval Committee Annual Report*, *IETS Newsl.*, 2005, **23**, 11-17.

LEVELS OF SOME NUTRIENTS AND NON-ESSENTIAL METALS IN COWS' WHOLE MILK FROM ADDIS ABABA, ETHIOPIA

E. Admasu,[1,*] B.S. Chandravanshi,[1,*] T. Wondimu,[1] and G. Akalu[2]

[1]Department of Chemistry, Addis Ababa University, Addis Ababa, P. O. Box 1176, Addis Ababa, Ethiopia, *E-mail: bhagnan@chem.aau.edu.et, or aengd@yahoo.com
[2]Ethiopian Health and Nutrition Research Institute, P. O. Box 1242, Addis Ababa, Ethiopia

1 INTRODUCTION

Diverse rainfall and temperature patterns are largely the result of Ethiopia's location in Africa's tropical zone and the country's varied topography. Winds that originate over the Atlantic Ocean and blow across Equatorial Africa have marked seasonal effect on much of Ethiopia. The resulting weather pattern provides the highlands with most of its rainfall during a period that generally lasts from mid-June to mid-September. The main rainy season is usually preceded in April and May by converging northeast and southeast winds that produce a brief period of light rains, known as balg. These rains are followed by a short period of hot dry weather, followed by violent thunderstorms occurring almost daily toward the middle of June.[1]

The arid, semi-arid and sub-humid zones are each homes for 14% of the cattle population, while 6% and 52% of the cattle population inhabit the humid zones and the highlands of the country, respectively. The majority of the cattle population is found in the highlands of Ethiopia where 43.6% of the human agricultural population is residing which indicates that cattle have a very important role in the Ethiopian economy.[2]

Animal products contain important nutrients in bioavailable form, essential for growth and proper physical and mental development.[3] Milk and milk products are the major source of essential nutrients in diet. Thus, knowledge of the mineral concentrations in milk samples and milk products is of particular interest. This becomes especially important in view of the wide variability in mineral content of milk due to numerous factors affecting milk composition, such as lactation (the composition of milk varies considerably during lactation, with the major changes usually occurring soon after the start of the lactation period), the breed of animal, climate, season, dietary composition of animal feed and soil contamination. Also, the mineral content may vary because of its handling by humans.[4]

An estimated total of 1.4 million metric tons of milk is produced in Ethiopia from an estimated population of 4.6 million cows, 9.3 million goats, 3 million ewes and 0.15 million camels. On a national scale, the per capita milk consumption is estimated at 19 to 30 kg per annum.[5] A recent survey undertaken by the Addis Ababa Agricultural Bureau shows that there are a total of 5,167 small, medium and large dairy farms in and around Addis Ababa city. The total milk production from these dairy farms amounts to 34,649,450 liters per annum. Of this, 73% is sold, 10% is left for household consumption, 9.4% goes to calves and 7.6% is processed, mainly into butter and cheese. The total amount of milk available to Addis Ababa is 43,849,675 liters per annum.[6] This figure is higher than the total dairy production from the dairy farms, the difference being accounted for by the informal milk supplied by the individual herds.

The study of mineral composition of whole milk becomes highly relevant especially to developing countries where malnutrition is a major threat to toddlers and the elderly. Moreover, data linking tropical topography to levels of nutrients in cows' whole milk are limited. Such studies assist in determining deficient nutrients and major causes of nutrient deficiency in whole milk. Although there exist a number of dairy farms in Ethiopia, data on mineral content of whole milk and factors affecting its composition are highly scarce. The objective of this research was (1) to determine essential and non-essential metals, fat and protein contents of milk samples; (2) to compare nutrients in pure Jersey breed milk (found only in one farm in the country), Holstein breed and the cross breed cows' milk; (3) to evaluate environmental influences on milk samples from various dairy farms; and (4) to compare metals, fat and protein contents of milk samples with literature reports.

2 MATERIALS AND METHODS

2.1 Sample collection

Eight cows' bulk milk samples were collected from four different dairy farms in Ethiopia, selected on the basis of their breed type, supply amount, and geographical location. Samples at all four farms were collected in two different periods of the year (December 2004 and February 2005) to better evaluate the influence of different feeding patterns on the levels of trace elements in the cows' whole milk. In this work, samples in December are for wet feeding season and the samples in February are for dry feed season where the average milk production level of the cows in each farm lowers significantly.

The number of cows represented in each bulk milk sample, location, altitude and breed reared for each farm are reported in Table 1. Bulk milk samples were collected at each farm. A total of 618 samples from four farms (twice from each 78 Jersey; 125 Holstein; 59 Holstein; and 47 cross breeds, respectively) were collected according to the Colorado Public Health and Environment milk sampling and testing manual 2000 and the International Dairy Federation (IDF) standard methods of sampling for bulk milk (tankers, silos, etc.). As recommended in these manuals, in the case of small retail samples, the whole container represents the sample, but must not be less than 100 ml.[7] The sample characteristics of cows' whole milk are described in Table 1.

Table 1 Background information on cows' whole milk sampling sites and number of cows sampled

Farm	No. of cows sampled	Farm location*		Altitude (m)	Breed
Adaberga	78	38°22' E, 9°18' N	Inchini area	2635	Jersey
Ayele	59	39°00' E, 8°45' N	Mojo	1781	Holstein
Selale No. II	125	38°52' E, 9°34' N	Mukaturi	2631	Holstein
Genesis	47	39°07' E, 8°36' N	Debrezeyt	1995	Cross breed

*Source: The Government of Ethiopia, Ministry of Land Reform and Administration, Survey and Mapping Department, 2004.

In particular, in each farm from each cow of different age, size and lactation period a minimum of up to 150 ml of milk was collected, mixed in the metal milking bucket and homogenized using vortex. The samples (750 ml) from each farm were transferred separately into low-density polyethylene (LDPE) bottles to 75 % of the volume of the bottle. In this way, four samples (one representative from each dairy farm) were brought to the laboratory in less than 24 h after milking.[7] Each milk sample was frozen at -20°C and then freeze-dried using a Labconco-3 freeze dryer. Triplicate analysis was then done on each sample. Thus a total of 4 x 3 = 12 samples were analyzed for the first period of sampling. Similarly 4 x 3 = 12 samples were analyzed from the second round samples. Standard reference material (SRM 8435: whole milk powder) was purchased from the National Institute of Standards and Technology (NIST), Gaithersburg, MD, USA.

2.2 Reagents

All dissolutions were made using analytically pure reagents. Stock standard solutions of 1000 mg/l (Buck scientific PURO-Graphic^{tm} calibration standards, prepared as nitrates for each element in 2% HNO_3) were used for preparing calibration standards. The calibration solutions were made using distilled and deionized water immediately before use. Nitric acid (70% trace metal grade; Spectrosol® BDH, England), perchloric acid (70% Analar® BDH, England), lanthanum nitrate hydrate (99.9% Aldrich, Milwaukee, USA), potassium sulfate (Analar® BDH, Poole, England), sulfuric acid (98% BDH, Poole, England), hydrogen peroxide (30% BDH, Poole, England), sodium hydroxide (98% Analar® BDH, Poole, England), boric acid (Anhydrous, 99%, Sigma grade, St Louis, USA), bromo cresol (Merck, Germany), methyl red (Fluka, Switzerland), ethanol (99% Absolute, BDH, Poole, England), and diethyl ether (Analar® BDH, Poole, England) were used as received.

2.3 Equipment

A BUCK SCIENTIFIC MODEL 210VGP (USA) atomic absorption spectrometer equipped with a deuterium arc background corrector was used for analysis of the analyte metals; a Kejeltec Auto 1003, Tecator, Hoganas (Sweden) was utilized for collecting ammonia from the digested sample in the analysis of protein; a Tecator/Udy, Ra, Fa Tec Extraction Unit, Hoganas (Sweden) was used for fat extraction; a Selecta model 2001241 (Spain) drying oven was used for drying glassware; a Freeze Dry-3, LABCONCO (USA) was used to freeze dry the whole milk samples; and a Kjeldahl apparatus was used as a hot plate during digestion of milk samples.

2.4 Quality control

Washing of sample containers, the digestion vessels, glassware for standards and sample tubes for metal determinations was made following recommended procedures.[8] In brief, sample containers and other glassware were cleaned with metal-free nonionic detergent solution, rinsed many times with tap water, soaked in 50% HNO_3 acid for 24 h and then rinsed three times with metal-free deionized water. All sample containers were rinsed again with deionized water prior to use.

2.5 Sample digestion and flame atomic absorption spectrometer analysis of samples

Aliquots ranging between 0.5 g and 1.0 g of freeze dried samples were placed in each of the four micro Kjeldahl (150 ml) vessels. The contents were digested on a micro Kjeldahl flask digestion unit (setting the temperature dial at 8) for 2.5 h after addition of 2.5 ml of conc. HNO_3 (70%) and 1.0 ml of conc. $HClO_4$ (70%). The contents were then further digested for an additional 2 h after introducing 1.0 ml of conc. HNO_3 and 0.5 ml $HClO_4$ into the digestion vessels until the solution was clear. After adding a solution of $La(NO_3)_3.xH_2O$ (to make the solution about 10 g/l expressed as La), the sample was transferred to a 25 ml volumetric flask and diluted with distilled deionized water to its mark. Levels of various metals in these solutions were determined using a flame atomic absorption spectrometer.

2.6 Protein

Milk was digested in H_2SO_4 using $CuSO_4.5H_2O$ as catalyst with K_2SO_4 as boiling point elevator, to release nitrogen from the protein and to retain nitrogen as ammonium salt. Concentrated NaOH was added to release NH_3, which was distilled, collected in H_3BO_3 solution and titrated with 0.10 M HCl.[9] Total protein was calculated as %total N x factor of 6.38.

2.7 Fat

Percent fat in freeze dried milk samples was determined by the Babcock method.[10] The fat content of the freeze dried milk was assayed by an ether extraction method. After extraction, aliquots of the lipid extract in triplicate were volatilized under an exhaust hood, dried in a laboratory oven at 105 °C for 2 h, and the fat contents of the freeze dried milk samples were quantified by determining percent dried weight relative to the original milk weight.

2.8 Statistical analysis

Differences among farms and between periods were analysed by means of ANOVA.[11]

3 RESULTS AND DISCUSSION

3.1 SRM 8435 analysis

To check the precision and accuracy of the method, a standard reference material (SRM 8435: whole milk powder) was analyzed. The results of the analysis are presented in Table 2. The results are generally in good agreement with NIST certified values, indicating the validity of our procedure for whole milk analysis.

3.2 Moisture determination

Analytical results compare best when the moisture content of milk samples is accurately determined. Water determination in freeze dried whole milk powder can be performed using either direct or indirect methods. The most common direct techniques are based either on desiccation or on oven drying.[12] In this research work, freeze drying followed by the oven drying method was used.

Table 2: *Results of the analysis of the whole milk reference material SRM 8435*

Analyte	Found (mg/kg)[a]	Certified (mg/kg)[b]
Calcium	0.85 ± 0.124*	0.92 ± 0.049*
Cadmium	nd	0.0002
Iron	2.10 ± 0.02	1.8 ± 1.1
Magnesium	784 ± 2.3	814 ± 76
Nickel	nd	0.01
Lead	0.10 ± 0.01	0.11 ± 0.05
Zinc	31.20 ± 0.01	28.00 ± 3.1
Protein	24.88 ± 0.7*	25.86 ± 0.7*
Fat	18.10 ± 2.6*	21.30 ± 2.4*

[a]Mean ± SD (n = 3). [b]Mean ± SD (n = 2); *results expressed in mass/mass, % m/m.
nd: not detected, n: number of the analyses carried out on the same milk reference sample.

This is a standard gravimetric method which consists of drying 1-3 g of a test sample at 102.0 ± 2.0 °C under atmospheric pressure for 2 h. Constancy of mass was tested by additional drying steps of 1 h until the difference in mass did not exceed 0.5 mg. The number of samples in the oven was limited to six to provide comparable amounts of released volatiles during all measurement series.[12] The results are shown in Figure 1.

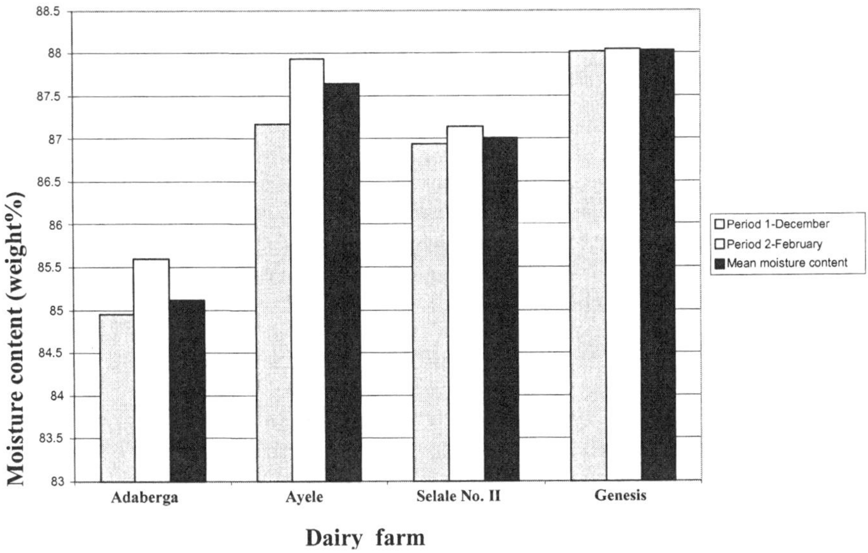

Figure 1: *Variation of moisture content determined using drying method IDF standard 26A in cows' whole milk collected from four dairy farms.*

The comparison of the moisture content of the milk of the two periods of sampling showed slightly higher moisture content values in period 2 (February 2005) than in period 1 (December 2004) for samples from farms Adaberga, Ayele, and Selale No. II. This may be attributed to the feeding habits of the cows which may be due to much consumption of water because of loss of consumed water by high respiration rates during this dry season and due to the dry feed. However, the moisture content from the Genesis farms remained constant, which may be explained by the fact that the cows feed on green matter throughout.

When the moisture content was compared between farms, the variation was pronounced in their breed type. Milk sample from the Jersey breed in Adaberga farm had the lowest moisture content, the Holsteins from Selale No. II and Ayele farms intermediate and the cross breed from the Genesis farm the highest. These results highlight the dependence of moisture content of cows' whole milk on such factors as feeding habits, breed type, and geographic locations. The results further signify the importance of moisture content determinations to be able to compare nutrient concentrations among milk samples obtained from different sources and locations.

3.3 Levels of essential and non-essential elements in cows' whole milk

3.3.1 The effect of sampling farms. The element levels in the four sampling farms (Adaberga, Ayele, Selale No. II and Genesis) were statistically evaluated to find out whether the observed differences in analytical results of samples collected from these farms are significant. Adaberga Farm is significantly different from the Ayele, Selale No. II and Genesis farms ($P = 0.05$) for Ca, Mg and Zn from the essential elements analyzed. But for Fe and the analyzed non-essential elements (Cd, Ni, Pb and Tl) there were no significant difference among the farms. Calcium, in fact, is a basic constituent of the casein complex and, thus, its content is directly related to protein content of milk. This could explain the difference observed between farms on calcium content. The milk from Jersey cows is rich in protein and fat compared to milk from other breeds. Consequently the higher content of calcium observed in Adaberga farm is probably related to its protein content (breed effect) and for the non-essential elements and Fe most likely due to the feedants and the processes involved in the milking, transport and storage factors that can affect the mineral concentrations in milk for human consumption.

It is important to note that milking was done manually into a metal bucket and one person was assigned to milk at least 10 cows one after the other. While the assignment of a single person to milk at least 10 cows a day may have negligible role in the levels of metals in the milk, collection in a metallic bucket may contribute metals to the milk.

3.3.2 Effect of sampling period. In general, when the effect of the sampling periods was investigated, results of variance analysis for whole milk (Table 3) showed highly significant differences between the two periods, at $P \leq 0.05$ levels for calcium, magnesium and protein in the analyzed samples. This is similar to literature results that reported higher calcium and magnesium values with significantly lower contents of zinc.[13] However, unlike the reported results, higher levels of iron were observed in the present study, and these are most likely associated with contamination during milking and the nature of their feeds; which in turn can be linked to contamination of the soil of grazing land by fertilizer residues. For non-essential metals (Table 3), results of variance analysis for cows' whole milk showed no significant differences between the two periods of sampling (December and February) at $P \leq 0.05$.

Table 3: *Mean concentration of essential and non essential elements in cows` whole milk samples collected in two sampling periods (n = 3)*

	Adaberga		Ayele		Selale No. II		Genesis	
Element (mg/L)	**Dec.**	**Feb.**	**Dec.**	**Feb.**	**Dec.**	**Feb.**	**Dec.**	**Feb.**
Ca	2237±310	1606±376	1711±297	1285±171	1900±239	1110±300	1766±259	1187±149
Mg	71.9±1.1	70.5±0.7	64.3±1.7	60.2±0.8	72.8±1.3	66.0 ±0.2	61.5±0.4	52.9±2.2
Fe	1.5±0.05	1.3±0.02	1.3±0.01	1.4±0.03	1.3±0.05	1.6±0.04	1.7±0.05	1.5±0.03
Zn	6.0±0.3	6.5±0.3	3.0±0.2	4.4±0.2	3.8±0.2	6.1±0.2	3.8±0.3	5.2±0.3
Cd	0.08±0.02	0.1±0.02	0.02±0.00	0.04±0.01	0.04±0.01	0.1±0.04	0.04±0.01	0.04±0.01
Ni	2.2±0.09	1.5±0.02	1.7±0.08	1.0±0.02	1.7±0.06	0.8 ±0.03	1.4±0.05	0.9±0.05
Pb	4.1±0.08	1.9 ±0.05	3.6±0.07	2.4±0.06	1.1±0.05	1.8±0.05	3.4±0.05	2.6±0.05
Tl	1.3±0.05	2.3±0.07	3.1±0.08	2.1±0.04	2.3±0.05	1.4±0.05	2.5±0.03	2.2±0.02

Mean ± SD; **Period 1:** December 2004**; Period 2:** February 2005. n: number of samples.
P value > 0.05 (farms); p value < 0.05 (periods); p value > 0.05 for Zn and Cd (periods).

3.4 Determination of protein content

The results of total protein determination are given in Table 4. The observed data on protein levels showed no significant difference ($P > 0.05$) among Adaberga, Ayele and Selale No. II farms, but the values for Genesis farm were substantially different. This is most likely because the three farms share identical feeding cultures, in contrast to the Genesis farm which supplies its own green feed throughout the year regardless of breed type. Due to the breed effect, Adaberga Farm has significantly higher content ($P = 0.05$) from the other farms studied.

Table 4: *Protein content of cows' whole milk (n = 3) obtained from four dairy farms*

Dairy Farm	**Protein content (% m/m) in dairy farm**			
	Adaberga	**Ayele**	**Selale No. II**	**Genesis**
Period 1: December	3.75	3.21	3.41	2.84
Period 2: February	3.43	3.07	3.12	2.73
Mean protein content (% m/m)*	3.59±0.23	3.14±0.09	3.26±0.21	2.78±0.08

Mean ± SD, *mass/mass; p value > 0.05 (periods); p value < 0.05 (farms)

The total protein for the first period of sampling (December 2004) is significantly higher at $P = 0.05$ for Adaberga, Ayele, and Selale No. II Farms but Genesis farm showed no significant difference compared to the second period of sampling (February 2005). This is most likely due to the variation of the cows' feed in the Adaberga, Ayele and Selale No. II Farms as compared to the consistency of feed in the Genesis farm.

The influence of nutrition of a cow on its milk protein has been variable. Improving the supply of methionine and lysine has resulted in an increase in the protein content of the milk. Dietary fat has consistently been associated with reductions in milk protein content and increases in milk protein yield. Thus, as the milk yield increases, the protein content gets smaller, but we would expect protein in milk at 70 lb to be 2.92 % and at 80 lb to be 2.89 %.[14] Accordingly, the milk yield in Genesis Farm as compared to the other farms studied was significantly high and so it showed lower percentage in the protein level as compared to the other studied farms. The high milk yield in Genesis Farm can be noted from the milking frequency (three times in 24 h). Therefore milk protein content is negatively correlated with milk yield.

3.5 Determination of fat

Accurate determination of fat in certain foods is difficult due to the binding of the fat by the matrix. Most methods used to determine fat in these difficult matrices include a pretreatment step to denature or destroy the physical structure of the matrix and allow greater accessibility to the fat by the hydrolysis method. Whole milk obtained in Adaberga Farm contained significantly more fat ($P \leq 0.05$) compared to milk from the other three farms, mainly because the cows were of the Jersey breed. Whole milk obtained in Selale No. II Farm contained more fat ($P \leq 0.001$) compared to Ayele and Selale No. II Farms. Although differences were found, some of these differences could be related to the random error in this type of analysis. Also, the processes involved in the milking, transport and storage are factors that can affect the mineral concentrations in milk for human consumption. Total fat content results determined by Soxlet extraction are given in Table 5.

Table 5: *Fat content determined in whole milk samples obtained from four dairy farms*

Sampling period	Fat content (% m/m) in dairy farm			
	Adaberga	**Ayele**	**Selale No. II**	**Genesis**
Period 1: December 2004	5.08±0.04	3.28±0.02	3.43±0.08	3.26±0.01
Period 2: February 2005	5.05±0.01	3.15±0.01	3.66±0.07	3.20±0.04
Mean fat content (% m/m)*	5.06±0.02	3.22±0.07	3.54±0.14	3.23±0.06

*Mean ± SD in %, *mass/mass; p-value > 0.05 (periods); p value < 0.001 (farms).

In period 1 (December 2004), whole milk had a higher protein ($P \leq 0.001$) than in period 2 in the four sampled farms, but the fat content showed variation from farm to farm (Table 5). Variations in milk composition, especially in the fat content, are dependent on the breed of the cattle. Also, the milking intervals significantly influence the fat content. For example, milking at an interval of 2 h leads to a fat content of 6%, compared to 3.6% fat obtained from milking at an interval of 12 h.[15]

3.6 Contributions of nutrients in milk to daily dietary intakes

To assess the nutritional status of the Addis Ababa population, evaluation of the contribution of milk to daily dietary intakes is compared to the recommended dietary allowance set by US National Research Institute (Table 6).

Table 6: *Contribution to daily dietary intake of Ca, Mg, Fe and Zn from the consumption of cows` milk in adult population of Addis Ababa, Ethiopia*

Element	RDA	Contribution of 60 mL of cows` milk[1]	
		Amount	**%RDA**
Ca (mg)	800	96.00	12.00
Mg (mg)	350 (300)[2]	3.90	1.11 (1.30)
Fe (mg)	10 (15)[2]	0.09	0.90 (0.60)
Zn (mg)	15 (12)[2]	0.29	1.93 (2.42)

[1]Source: Ethiopia Fact sheet: UNDP Human Development Report, 2002; calculated from milk consumption of 22 kg/capita.
[2]The recommended values for females are reported in parentheses.

The milk consumption in Ethiopia is reported to be 22 kg/capita,[16] which equates to 22 L of milk per 365 days per person (which is equivalent to 60 ml per day per person). This is used to calculate the values in Table 6.

The average contributions of each element from the consumption of 60 ml cows' milk have been calculated and are presented in Table 6. These mineral intakes were compared with the recommended dietary allowances (RDA) for adults established by the U.S. National Research Council. It is established that cows' milk is an excellent source of Ca. Ethiopian mean consumption of 60 ml of cows' milk supplies 12% of the RDM requirements of Ca and smaller amounts of Zn and Fe to the RDA values (Table 6) confirming that cows' milk is a very poor source of these elements.[4]

Tables 7, 8, and 9 show a selection of literature data describing concentrations of nutrients in cows' milk reported from other countries. The Ca concentration is similar to the values reported from Spain[4] and Italy[17] but significantly higher than from other countries shown in Table 8. The results of Fe obtained in this study are similar to data reported from Spain[18] and the USA,[19] significantly lower than reports from Turkey[20] and Cuba,[21] but higher than from the other countries listed in Table 7.

Table 7: *Concentration of Fe and Zn from different countries*

Country	Fe (mg/L)	Zn (mg/L)	n	Description	Reference
Ethiopia	1.5 ± 0.03	4.9 ± 0.25	618	Whole (two periods)	This work
Italy	0.65 [1]	3.82 [1]	__	Raw	22
	__	4.80 [1]	__	__	23
	__	3.58 [1]	__	__	24
Spain	__	3.0 ± 0.2	9	Bottled cow milk	25
	__	3.1 [1]	70	Raw	26
	1.36 ± 1.65 [1]	3.78 ± 0.76 [1]	129	Raw	27
	0.14 [1]	3.82 [1]	300	Pasteurized	18
	0.44 ± 0.11 [1]	4.21 ± 0.41 [1]	120	Raw (12 months)	28
	0.44 ± 0.12 [1]	4.21 ± 0.41 [1]	120	Raw (12 months)	29
	0.46 ± 0.10 [1]	3.70 ± 0.22 [1]	10	Raw	30
	0.30 ± 0.03 [1]	3.63 ± 0.06 [1]	10	Pasteurized	
	0.29 ± 0.11	3.42 ± 0.02	27	Raw	31
	0.52 ± 0.18	4.41 ± 0.67	151	Raw (12 months)	4
	0.17 ± 0.03	3.06 ± 0.14	18	Sterilized	
Turkey	2.68 ± 0.24	3.64 ± 0.16	18	__	20
U.S.A.	__	(3 - 5)	__	Mature	32
	0.506	4.23	36	__	19
	0.20 ± 0.40 [1]	3.7 ± 0.7 [1]	__	__	33
	0.20 ± 0.30[1]	__	__	__	34
Cuba	2.9 ± 0.5	__	16	Pasteurized	21
Japan	0.27 [1]	__	__	Raw	35
Pakistan	0.60 [1]	4.20 [1]	__	Raw	36
Burundi	__	4.4 ± 0.3	19	Raw	37
		4.1 ± 0.3	2	Processed	
Egypt	0.15 [2]	3.46 [2]	5	Raw	38

[1] (mg/kg).
[2] (% total solids, TS = 11.54 %).

The concentration of Zn obtained in this study is higher than most of the reported values shown in Table 7, which could be associated with the level of copper in their feeding that interferes with the copper absorption system. It has been reported that as the content of copper increases the content of zinc decreases.[39] This conclusion implies that the levels of copper in our milk samples are likely to be low compared to those from other regions.

The moisture content, total protein and total fat contents given in Table 9 are similar to the reports from South Africa in different localities,[13] but reports from Portugal show significantly lower total fat values than that from the Barrosã breed[15] because of their milking duration compared in Table 9.

Table 8: *Concentration of Ca and Mg from different countries*

Country	Ca (mg/L)	Mg (mg/L)	n	Description	Reference
Ethiopia	1600 ± 263	65 ± 1	618	Whole (two periods)	This work
Spain	1060 ± 50 [1]	__	89	Raw (10 months)	40
	1089 ± 243	146 ± 20	129	Raw (12 months)	27
	1280 [1]	110 [1]	300	Pasteurized	18
	1199 ± 106 [1]	105± 7.5 [1]	120	Raw (12 months)	29
	1251 ± 48	116 ± 3 [1]	10	Raw	30
	1289 ± 60 [1]	115 ± 5 [1]	10	Pasteurized	
	1653 ± 207	114 ± 19 5	1306	Raw (12 months)	4
	1309 ± 62	121 ± 14	6	Sterilized	
Italy	1673	106	__	Whole	17
Turkey	__	92.3 ± 2.2	18	__	20
U.S.A.	1060 ± 280 [1]	98 ± 19 [1]	57	Whole	41
	1010 ± 200 [1]	100 ± 10 [1]	__	Whole	33
	0.128[2]	61	36	>3 months	32
Egypt	__	9.8 ± 0.3 [3]	5	__	38
South Africa	121.0 [2]	11.43 [2]	__	Whole (Summer)	13
	119.4 [2]	11.80 [2]	__	Whole (Winter)	

[1] (mg/kg); [2] %; [3]mg/g ash; n: number of samples.

Table 9: *The mean proximate composition of whole milk obtained in different countries*

Country	Composition (%m/m)			Breed	n	Reference
	Moisture (Weight %)	Protein (Weight %)	Total fat (Weight %)			
Ethiopia	85.11 ± 0.34	3.59 ± 0.22	5.06 ± 0.01	Jersey	156	This work
	87.65 ± 0.38	3.14 ± 0.09	3.22 ± 0.07	Holstein (Selale No. II farm)	250	
	87.01 ± 0.11	3.26 ± 0.20	3.54 ± 0.14	Holstein (Ayele farm)	118	
	88.03 ± 0.27	2.78 ± 0.07	3.23 ± 0.04	Cross breed	94	
South Africa	87.98	3.27	3.45	Locality	__	13
	87.69	3.28	3.56	Locality	__	
	87.95	3.20	3.46	Locality	__	
	88.31	3.19	3.29	Locality	__	
	87.96	3.30	3.38	Locality	__	
Portugal	__	3.27 ± 0.47	4.94 ± 1.44	Frísia	5	15
	__	3.96 ± 0.50	2.76 ± 0.43	Barrosã	5	

Table 10 shows literature data describing the levels of Pb and Cd in milk from different countries. In general, significantly higher concentrations of Pb are observed in this work compared to reports from Germany[39] and India.[42] The presence of relatively higher levels of Pb in the Ethiopian milk samples could be due to various factors, such as transport vehicle emissions along roads and/or motor ways, fodder contamination, climatic factors, such as winds, and the use of pesticide compounds. The concentration of Cd is found to be comparable to that reported from Germany[41] and India.[42] Therefore, it is necessary to monitor the levels of these metals, especially of Pb, from time to time to better clarify their presence in milk.

Table 10 *Concentration of Pb and Cd in cows' milk from different countries*

Country	Pb	Cd	n	Description	Reference
Ethiopia (mg/L)	2.6	0.06	618	Whole milk	This work
Germany (μg/L)	1.8	0.1	__	Cow milk	42
India (μg/L)	1.7	0.07	75	Cow milk	43

4 CONCLUSIONS

The contents of essential minerals (Ca, Mg, Fe and Zn), protein and fat of cows' whole milk distributed in Addis Ababa fall into normal levels described in the literature, though the non-essential elements (cadmium, nickel, lead and thallium) show significantly higher levels. The average consumption of cows' milk distributed in Addis Ababa supplies 12% of the RDA requirements of calcium and small amounts of magnesium and zinc. Cows' milk sampled in December presented higher concentrations of calcium, magnesium and iron, indicating higher contributions of these elements to the consumer in milk collected at this time than in milk sampled in February.

References

1 Federal Research Division of the Library of Congress (1986 and 1998) as part of the Country Studies/Area Handbook Series, U.S. Library of Congress Available at: http://countrystudies.us/ethiopia/41.htm; Accessed 17/02/2006.

2 H.E. Jahnke, 1982. Livestock production systems and livestock development in Tropical Africa. Pp 253. Federal Republic of Germany: Kieler Wissenchaffsverlag Vauk.

3 M. Yishak, S. Ehui, M. Jabbar and B. Shapiro, 1999. Handbook of Livestock Statistics for Developing Countries, ILRI: Addis Ababa, Ethiopia.

4 E.M. Rodríguez Rodríguez, M. Sanz Alaejos, and C. Díaz Romero, *J. of Food Composition and Analysis*, 2001, **14**, 419.

5 L.R. Kurwijila, 1990. Nutritional importance of traditional milk products in the national diet. 1. Southern and Eastern Africa general. Available at: http://www.fao.org/docrep/003/t0251e/T0251E01.htm; Accessed 02/05/2005.

6 T. Azage, T. Million, M. Yoseph, and Y. Alemu, 1998. Market-oriented Urban and Peri-Urban Dairy Systems, ILRI: Debre Zeit Research Station, Ethiopia.

7 G. Sahin, A. Aydin, A. Isimer, I. Ozalp and S. Duru, *Biological Trace Element Research*, 1995, **50**, 87.

8 American Public Health Association, 1998. Standard Methods for the Examination of Water and Wastewater, 20th edn. New York, USA: American Public Health Association, American Water Works Association and Water Environment Federation.
9 D.M. Barbano, J.L. Clark, C.E. Dunham and J.R. Fleming, *J. of the Association of Official Analytical Chemists*, 1990, **73**, 849.
10 G.H. Richardson, 1985. In: Standard Methods for the Examination of Dairy Products, 15th edn., p. 347, Washington, DC, USA: American Public Health Association.
11 E.D. McBean and F.A. Rovers, 1998. Statistical procedures for analysis of environmental monitoring data & risk assessment. New Jersey, USA: Prentice Hall.
12 C. Reh, S.N. Bhat and S. Berrut, *Food Chemistry*, 2004, **86**, 457.
13 L.E. Smit, H.C. Schönfeldt, W.H. J. de Beer and M.F. Smith, *J. of Food Composition and Analysis*, 2000, **13**, 345.
14 D.F. James, 2004. Milk Protein Center for Animal Health and Productivity, School of Veterinary Medicine, University of Pennsylvania, Available at: http://cahpwww.vet.upenn.edu/mun/milk_protein.html; Accessed 02/05/2005.
15 P. Pires, É. Fernandes, M. Vilarinho, M. Barros, R. Ferreira, L. Careiro, G. Almeida, M. vaz Velho, *Electronic Journal of Environment*, 2003, **2**, 1.
16 Ethiopia fact sheet: UNDP Human Development Report, 2002.
17 D. Mazzotta, V. Brandolini, D. Vecchiati, E. Menziani, A. Mallea Angles, F.S. Pansini and V. Abbasciano, *Rivista Science Alimentación*, 1993, **22**, 287.
18 R. Moreno-Rojas, M.A. Amaro-LÓpez, and G. Zurera-Cosano, *Rev. Esp. Cienc. Tecnol. Aliment*, 1993b, **33**, 435.
19 Y. Salih, L.R. McDowell, J.F. Hentges, R.M. Mason, and C.J. Wilcox, *J. of Dairy Science*, 1987, **70**, 608.
20 A. Yenigün, B. Taneli, and F. Oksel, 1995. Trace toxic elements in nutrition and health. In: Proceeding International Conference on Health and Diseases: Effects of Essential Toxic Trace Elements, 4th edn. Pp. 339-340.
21 L. Batilde Lima, S. Banguela Perez, M.J. De Ortega Bolaños, R. Torricella and J. Camejo Corrales, *Aliment*, 1995, **259**, 63.
22 V. Del Petre, and F. Di Stanislao, *Nuovi annali d'igiene e microbiologia*, 1984, **35**, 313-323.
23 M. Baldini, A. Bocca, and M. Mosca, *Food Additives and Contaminants*, 1988, **5**, 45.
24 M.A. Franco, F. Balestrieri, M. Sabbatini and A. Serra, *Rivista di Scienza dell'Alimentazione*, 1981, **10**, 35.
25 N. Campillo, P. Viñas, I. López-García, and M. Hernández-Córdoba, *Talanta*, 1998, **48**, 615.
26 J.E. Barreira Rodriguez, and S. Arribas Jimeno, *Anales de Quimica International Edition*, 1990, **86**, 252.
27 P. López Mahía, P. Paseiro Losada and J. Simal Lozano, *Aliment*, 1991, **226**, 45.
28 R. Moreno-Rojas, M. A. Amaro-López, and G. Zurera-Cosano, *International J. of Food Sciences and Nutrition*, 1993a, **44**, 37.
29 F. Rincón, R. Moreno, G. Zurera and M. Amaro, *J. of Dairy Research*, 1994, **61**, 151.
30 G. Zurera-Cosano, R. Moreno-Rojas and M. Amaro-López, *Food Chemistry*, 1994, **51**, 75.
31 C. Martín-Hernández, L. Amigo, P.J. Martín-Alvarez and M. Júarez, *Deutsche Forschungsanstalt für Lebensmittelchemie*, 1992, **194**, 151.
32 B. Lönnerdal, C. L. Keen and L. S. Hurley, *Annual Review of Nutrition*, 1981, **1**, 149.
33 J.A.T. Pennington, S.A. Schoen, G.D. Salmon, B. Young, R.D. Johnson and R.W.J.E. Marts, *J. of Food Composition Analysis*, 1995b, **8**, 129.

34 J.A.T. Pennington and B.E. Young, *J. of Food Composition Analysis*, 1990, **3**, 166.
35 H. Gunshin, M. Yoshikawa, T. Doudou and N. Kato, *Agricultural and Biological Chemistry*, 1985, **49**, 21.
36 N. Bano, M. Naeem and H.H. Khan, *J. of Natural Sciences and Mathematics*, 1985, **25**, 67.
37 H. Benemariya, H. Robberecht and H. Deelstra, *Science of the Total Environment*, 1993, **128**, 83.
38 M.N. Rashed, *International J. of Environmental Analytical Chemistry*, 1992, **48**, 41.
39 P. Licata, D. Trombetta, M. Cristani, F. Giofre, D. Martino and M. Calo, *Environmental International*, 2004, **30**, 1.
40 M.R. Fernández De La Campa, E. Andrés García, R. Moro García, M.L. Alvarez Bartolomé and S. Arribas Jimeno, *Aliment*, 1986, **23**, 33.
41 J.A.T. Pennington, D.B. Wilson, B.E. Young, R.D. Johnson and J.E. Vanderveen, *J. of American Dietary Association*, 1987, **87**, 1036.
42 P. Ostapczuk, R. Valenta, H. Rutzel and H.W. Nurnberg, *Science of the Total Environment*, 1987, **60**, 1.
43 R.M. Tripathi, R. Raghunath, V.N. Sastry and T.M. Krishnamoorthy, *Science of the Total Environment*, 1999, **227**, 229.

WATER, WATER EVERYWHERE AND WHEN FOUND NOT FIT TO DRINK

S.O. Wandiga

University of Nairobi, College of Biological and Physical Sciences, School of Physical Sciences, Department of Chemistry, P.O. Box 30197-00100, Nairobi, Kenya

1 INTRODUCTION

At no time has the world faced such an acute shortage of potable water than today. This is despite the fact that 75% of the world's surface is covered with water. Unfortunately, 97% of this water is saline and therefore undrinkable. Of the available fresh water, 2% is locked up in ice caps and glaciers. Hence there is only about 1% of water available to meet human needs.

Further analysis of the available water shows that the blue water locked up in lakes and reservoirs and green water-rainfall water have become increasingly polluted with human, industrial and agricultural wastes and cosmetic chemicals. Most of Africa's largest rivers are severely fragmented by dams, diversions and canals, leading to the degradation of ecosystems. Rapid urbanization in municipalities adds increased pressure on Water Quality.

Water-related diseases have shown an increasing trend. For example, 72% of all reported cholera cases in the world in 1998 were in Africa. Poor water quality is a key cause of poverty. About 3.1 million people died in 2002 as a result of diarrheal diseases and malaria, 90% of who were children.

Water quality is declining in most African regions, affecting the diversity of freshwater species and ecosystems. There has been a marked decline of freshwater resources due to bad land use and management practices. In the last 20 years, countries such as Niger lost more than 80% of their freshwater wetlands. More than 600 lakes in Africa have shrunk dramatically over the past decades, drained by deforestation, pollution and farming. Many urban water supply systems are "losing" 30-40% of water through leakage and illegal extraction. Due to population pressure, the world will require 55% more food (mainly for Africa) by 2030, increasing the demand for irrigation.

Evaporation and ground filtration are the two natural processes for water purification. However, these are not sufficient technologies to meet the challenges of potable water for human consumption. Scientific challenges exist in determining the quantities of such wastes and their breakdown products, effects on life and environment and how best to control their distribution in the environment. This paper describes some of the challenges existing in the East Africa sub-region.

1.1 Climate change in Eastern Africa affects water availability

Climate change and climate variability have added to the urgency of the need for water conservation. This is because the frequency of drought and flooding episodes has increased and climate change has altered the temperature and rainfall patterns in Eastern Africa and the continent. The temperature scenario for Africa described by Hulme *et al.*[1] shows that global mean surface temperature is projected to increase by between 1.5°C (2.7°F) and 5.8°C (10.8°F) by 2100; climate change scenarios for Africa indicate future warming across the continent ranging from 0.2°C (0.36°F) per decade (low scenario) to more than 0.5°C (0.9°F) per decade (high scenario); and this warming will be greatest over the interior of semi-arid margins of the Sahara and central Southern Africa.[1] The warming trend is expected in all seasons.

The fourth Intergovernmental Panel on Climate Change (IPCC) (AR4), 2007, concludes that "changes in climate are now affecting physical and biological systems on every continent.[2] Effects on human systems, although more difficult to discern due to adaptation and non-climatic drivers, are emerging. Over 90% of observed changes in systems and sectors are consistent with regional temperature trends. Many of the changes are now attributed to temperature increase caused by anthropogenic greenhouse gas emissions". A warmer climate increases the evaporation of water as well.

Rainfall is similarly affected under climate change scenarios.[1] The trend shows a decline of rainfall in Southern Africa but some probable increase in the Sahel and East Africa. However the quantity of rainfall will vary even within a country.

Sub-Saharan Africa is likely to experience large impacts from climate change, due to high levels of intrinsic vulnerability stemming from heavy dependence on rain-fed agriculture, propensity to drought, and relative low levels of adaptive capacity. The proportion of the African population at risk of water scarcity could increase from 47% in 2000 to 65% in 2025, when about 370 million African people may experience increased water stress based on a range of scenarios as observed by the IPCC.[2] Non-climatic changes such as water policy and management practice may add significant effects. Changes in primary production of large lakes will have important impacts on local food supplies. For example, Lake Tanganyika currently provides 25-40% of animal protein intake for the population of the surrounding countries and, on the basis of observed and paleoclimatic records, it is expected that climate change will reduce catches by 30%.[2] East African countries already have signs of water scarcity due to various reasons.

Some of the basic problems with water as a resource in Africa include: the very high potential evaporation, which is in excess of 2000 mm p.a. over large tracts and occurs throughout the year, very high aridity indices, a generally low ratio of conversion of rainfall to runoff, an often very concentrated seasonality of rainfall, and hence runoff, a strong response to the El Niño Southern Oscillation (ENSO) signal and thus generally high inter-annual coefficient of variability of rainfall, and an amplification of the inter-annual coefficient of variability of rainfall by the hydrological cycle.[3]

1.2 Drastic Water Quality Changes in East Africa

The quality of blue water in East Africa has experienced drastic changes in the last forty years as evidenced by the onset of massive cyanobacteria blooms offshore that took place in the Lake Victoria. The poor water quality has led to the collapse of the indigenous fish stock. The collapse has also been correlated to the poor agricultural practices that has led to an accelerated rate of deposition and sedimentation of nutrient-rich soils.[4] On top of activities related to human population and agricultural production is added the human abuse of water systems. Today most of our lakes and rivers are choking with wastes that are wantonly dumped into them. It is possible to prevent water pollution through enactment of stringent laws and regulations, public education and enforcement of laws and regulations. Investment in water cleaning technologies will greatly assist in water purification.

2 CHALLENGES FACING LAKE VICTORIA

Several challenges face Lake Victoria. Some of the obvious examples include soil erosion and the nutrient load that accompanies it, municipal and rural waste disposal of untreated sewage, either directly into the lake or through the rivers, effluents containing pesticides and fertilizers used in agricultural farms and leached into the water system, industrial and household waste dumped into the lake, rapid deforestation of water catchment forests and destruction of wetlands habitat, increasing fishing catch and fishery, urbanisation along the riparian areas, and rapid human and animal population growth.

The sediment load affects the lakes differently depending on their depth. Lakes Malawi and Tanganyika are deep lakes and the effect of siltation is not currently observed. However, for a shallow lake like Lake Victoria, siltation has reduced the depth from about 80m in the deepest area to an average of 40m today. The lake is receding and this may be attributed to some extent to siltation and climate change. In addition to the lake's depth, the bottom oxygen content and transparency (the Secchi index) decreased from 5m in 1930 to less than 1m in 1990s), while sediment and water phosphorus and nitrogen concentrations have also increased.[5] These changes have resulted in:

- Increased soil erosion and vulnerability to floods and droughts;
- Loss of soil fertility, declining crop yields;
- Food insecurity;
- Loss of dry season gardening, livestock husbandry capability and groundwater recharge and discharge due to degradation of floodplain wetlands.

One of the scientific challenges facing scientists in equatorial Africa is to ascertain the source of phosphorus that causes eutrophication in the equatorial lakes. The Lake Victoria Environmental Management Project[6,7] estimated the nutrients input into the lake with atmospheric deposition contributing 102,000 t/y of Total-N and 24,000 t/y of Total-P. These values indicated that the atmospheric deposition is by far the most significant contributor to the overall nutrient budget of the lake. Several authors have given various estimates as shown in Table 1.

Table 1 *Various Estimates of Amounts of Atmospheric Deposition of phosphorus*

Author(s)	Atmosphere
Reference 8	55%
Reference 9	90%
Reference 10	63.7%

As Table 1 indicates, the atmospheric phosphorus deposition debate has been with us for the last ten years. The fact that there are no phosphorus mines within the Lake Victoria catchments points to the possible largest source of atmospheric phosphorus coming from human activities such as biomass burning. However, evidence of large scale biomass burning is not there. Further, given the fact that the wind regime around the lake is influenced by NW and SW monsoons, it is doubtful that these estimates give a clear picture of the source of the atmospheric phosphorus load. A new project is being launched under the Global Environment Facility (GEF), UNEP/World Bank execution to explore the origin of phosphorus in equatorial lakes. In the meantime, our studies have started to shed light on the sources of phosphorus in the waters and sediments of Lake Victoria.

One of the major sources of water pollution in East Africa is human waste. The effluents from untreated municipal sewers pose great danger to the conservation of a sustainable ecology in and around Lake Victoria. Municipal sewage contains both faeces and urine that are sources of phosphorus. Therefore let us make two assumptions that will enable us calculate the phosphorus contributions from these sources. Let us assume that each person produces 25-50 kg/yr of faeces which contain 0.18 kg P; second, assume that each adult produces about 400 liters of urine per year, depending on liquid consumption, and contains 0.40 kg P. This is because municipal and industrial wastewater treatment plants are known to be the major point sources of phosphorus in urban areas.[11] Waste disposal sites, construction sites, fertilizers and farmyards also make substantial contribution to the total phosphorus load.[12-15] However, all these have not been adequately evaluated. Given the number of sewered and unsewered municipalities and their populations in Table 2, one is able to calculate the amount of phosphorus produced.

Table 2 *The population of sewered and unsewered municipalities in the Lake Victoria catchments*

Country	Total population (1000 people)	Sewered Urban population (x1000)	Unsewered Urban population (x1000)	Number of towns
Kenya	10,200	390	630	18
Uganda	5,600	210	870	9
Tanzania	5,200	27	340	4
Rwanda	5,900	-	400	5
Burundi	2,800	-	140	4
Total	**29,700**	**627**	**2,380**	**40**

Using these figures it is possible to calculate the phosphorus load coming from this source as given in Table 3. If we assume that the sewered treatment plants are working efficiently, an assumption we know is not true, then the P contribution from unsewered municipalities is about 397,000 t/y, a figure which is much larger than that reported in the Lake Victoria Environmental Management Program (LVEMP) reports.[6,7]

Table 3 *Calculated phosphorus discharge into the lake from sewered and unsewered municipalities*

	Faeces				Urine
Country	Sewered Pop.*	Phosphorus t/y	Unsewered pop*	Phosphorus t/y	Phosphorus (sewered-unsewered) t/y
Kenya	390	1,755-3,510	630	2,835-5,670	61,400-100,800
Uganda	210	945-1,890	870	3,915-7,830	33,600-139,200
Tanzania	27	122-243	340	1,530-3,060	4,320-54,400
Rwanda	-	-	400	1,800-3,600	- 64,000
Burundi	-	-	140	630-1,260	- 22,400
Total	627	2,822-5,643	2,380	10,710-21,420	99,320-380,800

*Pop=Population (x 1000)

If we assume that P contribution is from sewered municipalities with inefficient operations, then we have even a larger figure of approximately 500,000 t/y. On the other hand, if we assume that only 20% of P from this source reaches the lake, then we calculate a figure of 79,000-100,000 t/y. However, it is not possible that only 20% of P nutrient from this source will reach the lake, but even this assumption gives a figure that is still larger than that found in the LVEMP reports.

We have followed this theoretical calculation of P content with measurements of the element in water, soil and sediments. Water, soil and sediments were sampled for two years covering four different seasons per year (wet, two dry and short rain seasons) experienced in the Kenyan Lake Victoria catchments. Sampling points covered lakeshores, river mouths and effluent discharge points and the parameters analysed included total reactive phosphates, total hydrolysable phosphates, total phosphate, sediment exchangeable phosphates, sediment bio-available phosphate and soil-available phosphates. Soils from the catchments were found to contain ten to 100 times higher concentration compared to sediments and water samples. Water from both the rivers and the lake were found to contain phosphate levels much higher than the recommended guidelines for aquatic life, indicating the influence of anthropogenic sources.[16]

The seasonal average of total phosphate in the water were 4.61, 3.43, 2.45 and 2.30 mg/l for wet, short rain and dry seasons 1 and 2 respectively, whereas the total reactive phosphates had means of 2.22, 2.08, 1.12 and 1.61 mg/l in the same seasons. Sediment bio-available phosphates were higher than exchangeable phosphates with the highest mean concentrations of 24.45 and 8.22 mg/kg occurring during the dry season, whereas the average soil-available phosphorus ranged between 639 and 1076 mg/kg. Pearson correlation analysis of the data indicated a strong positive correlation between the levels detected in the water for all the seasons, implying increasing accumulation of the nutrients in the drainage basin (Table 4).

Table 4 *Pearson correlation analysis for phosphates in water*

Correlations					
		WET SEASON	DRY SEASON 1	SHORT RAIN	DRY SEASON 2
WET SEASON	Pearson correlation	1.000	.938”	.966”	.966”
	Sig. (2-tailed)	.	.000	.000	.000
	N	32	32	32	32
DRY SEASON 1	Pearson correlation	.938”	1.000	.926”	.873”
	Sig. (2-tailed)	.000	.	.000	.000
	N	32	32	32	32
SHORT RAIN	Pearson correlation	.966”	.926”	1.000	.969
	Sig. (2-tailed)	.000	.000	.	.000
	N	32	32	32	32
DRY SEASON 2	Pearson correlation	.966”	.873”	.969”	1.000
	Sig. (2-tailed)	.000	.000	.000	.
	N	32	32	32	32
“. Correlation is significant at the 0.01 level (2-tailed)					

3 METHOD AND RESULTS

3.1 Analysis of phosphates

Total reactive phosphate in the samples was determined in triplicates. To 50 ml samples 0.05 ml phenolphthalein indicator was added and the red color was discharged by adding drops of 50% hydrochloric acid solution before diluting to 100 ml. Excess coloration was removed by shaking the sample with 200 mg activated carbon in a 250 ml Erlenmeyer flask for 5 minutes and filtering through Whatman filter paper number 42. All the standards and the blanks were treated in the same manner to eliminate the interferences due to carbon. Color development in the sample was achieved by transferring 35 ml of the sample filtrate into a 50 ml volumetric flask and adding 10 ml of vanadate-molybdate reagent. The mixture was topped to the mark with distilled water and thoroughly mixed. Absorbance was taken at 470 nm using a Shimadzu 1700 UV-Visible spectrophotometer for determination of total reactive phosphates.

Total hydrolysable phosphate in water samples was determined after preliminary hydrolysis with sulphuric acid solution. To 50 ml samples was added 0.05 ml phenolphthalein indicator and, in cases where a red colour was observed, the color was discharged with drops of 30 % sulphuric acid solution. To each sample, 1 ml of 30 % sulphuric acid solution was added and diluted to 100ml. The samples were boiled for at least 90 minutes on a hot plate and the final digest was reduced to 25 to 50 ml. To each digest, 6 N sodium hydroxide solution was added until a faint pink color was observed. The faint pink colour was then discharged by adding 50% hydrochloric acid solution and the sample then proceeded to determination of the phosphate using vanadate-molybdate method as in the case of total reactive phosphate. Both the standards and blanks for determination of total hydrolysable phosphate were treated in the same manner as the samples.

Determination of the total phosphate in the water samples was conducted using the persulphate method. To 50 ml of the sample was added 0.05 ml phenolphthalein indicator and the red colour was discharged by adding drops of 30 % sulphuric acid solution. The sample was then diluted to 100 ml and transferred into a 250 ml Erlenmeyer flask. To each sample was added 0.5 g potassium persulphate and the sample was boiled on the hot plate for 90 minutes. The digest was treated with 6N sodium hydroxide solution until it became faint pink in colour, which was then discharged by adding drops of 50% hydrochloric acid solution before diluting to 100ml. 1 ml of 1:1 hydrochloric acid solution was added to all the samples and total phosphate determination was conducted using the vanadomolybdophosphoric acid calorimetric method.

3.1.1 Analysis of polychlorinated biphenyls. We have undertaken analysis of PCBs along the shores of Lake Victoria. Sampling stations were chosen to capture upstream activities, run offs and sediments brought along the rivers as source of PCBs. Surface sediments from three different composite sites were sampled with a stainless steel shovel and combined to one pole sediment sample because PCBs in sediments can show a patchy distribution. Sediment samples were wrapped with aluminum foil, labeled and stored at -25°C.

3.1.2 Analysis of pesticides. The field samples were collected from 12 points located on River Nzoia, River Sio and Lake Victoria. Field sampling was done during the short rain (October-December, 2002), dry (January-March, 2003) and wet (April-June, 2003) seasons. Water was sampled by the grab method into 2.5 L amber bottles and preserved with 1 g mercuric chloride to stop microbial degradation of pesticides. In the laboratory, samples were stored in a refrigerator at 4^0 C prior to extraction. Extraction was done using

2.0 litres of water for all the samples. The water was neutralized to pH 7 using a phosphate buffer, whereas 100g sodium chloride was added to salt out the pesticides from the aqueous phase. In all cases extraction was achieved by extracting the water thrice using 60 ml portions of triple distilled dichloromethane. The extracts were combined and cleaned by eluting through a 20 g Florisil column with 200 ml 6%, then 15% and 50% diethyl ether: hexane mixtures.

Sediments were sampled using a pre-cleaned stainless steel shovel and mixed on clean aluminium foil. Three representative samples, each of approximately 500 g, were taken and wrapped in aluminium foil, packed in black polythene bags and placed in an icebox containing wet ice. In the laboratory, sediment samples were stored at -20^0 C in a deep freezer prior to extraction. Extraction was done in triplicates using 25 g sample sizes following the AOAC method 970.52 for multi-residue analysis of pesticides. Sample extracts were cleaned by eluting through the Florisil column as described for water samples above.

All cleaned water and sediment samples were reduced to 2 ml using a LABCON rotary evaporator and analyzed by Varian CP 3800 gas chromatograph equipped with electron capture detector and CP-SIL 8CB capillary column with dimensions 15 m x 0.25 mm x 0.25 μm film thickness using a temperature programme 150^0 C with zero hold time, then increased to 240^0 C at 4^0 C/minute and hold time of 4.5 minutes. Confirmation of pesticide residues was done using a DB 1701 capillary column with dimensions 15m x 0.53mm x 0.5 μm film thickness, using a temperature programme 150^0 C with zero hold time, then increased to 200^0 C at 4^0 C/ minute and hold time of 4.5 minutes. Identification and quantification of residues was done using high purity pesticide standards obtained from Dr Ehrenstorfer GmbH (Augsburg, Germany). The methods used had acceptable recovery rates ranging from 70 to 130%, as established by spiking blank sediment and distilled water samples and following the same extraction, cleanup and analysis procedures as for the real samples.

3.2 Available phosphates

The levels of bioavailability of phosphates in water, sediment and soil are shown in Figures 1, 2 and 3 respectively. The highest concentrations were observed at municipalities which have no sewerage treatments. Secondly, phosphate levels in the Lake Victoria drainage system vary with seasonal changes. The levels in soils along the shore are higher than at river mouth and the sites where there is direct sewage discharge are leading in concentration of phosphates.

Furthermore, due to high nutrient load at points of direct sewage discharge, it was found that macrophytes like water hyacinth flourish best near shore. Several sites, like the Homa Bay, Hippo Point and River Kisati, receive high loads of untreated sewage and industrial wastes. Water hyacinth also flourishes at the off shore of these sites.

3.2.1 *Polychlorinated Biphenyls.* Polychlorinated biphenyls (PCBs) are compounds derived from substitution of one to ten hydrogen atoms of biphenyl by chlorine atoms. PCBs have been produced and used world-wide in large quantities for many years as transformer oils, metal-cutting oils, hydraulic oils, heat transfer fluids, and as additives in plastics, dyes and carbonless copying paper. The production of PCBs was terminated world-wide around the late 1970s to early 1980s due to their adverse effects on the environment as a result of persistency, bioaccumulation properties and toxicity.

The amount of PCBs in the global environment has been estimated to be about 3.7×10^8 kg and a further 7.8×10^8 kg were estimated to be still available for utilization or

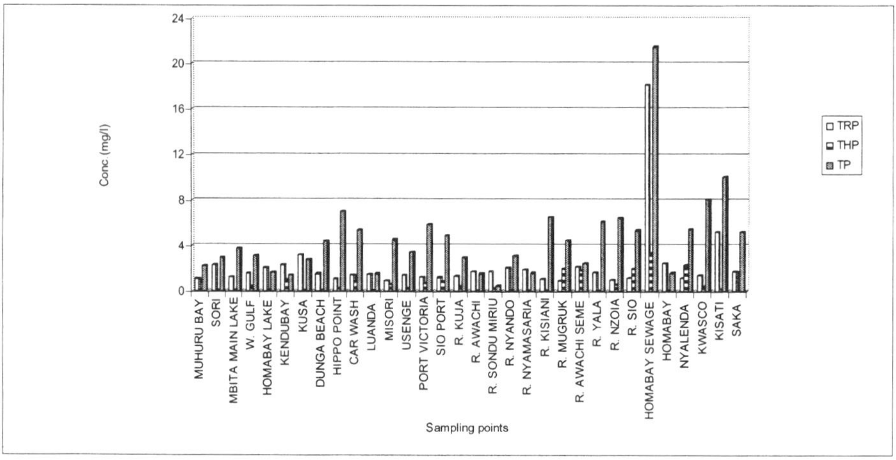

Figure 1 *Phosphates in water during wet season*

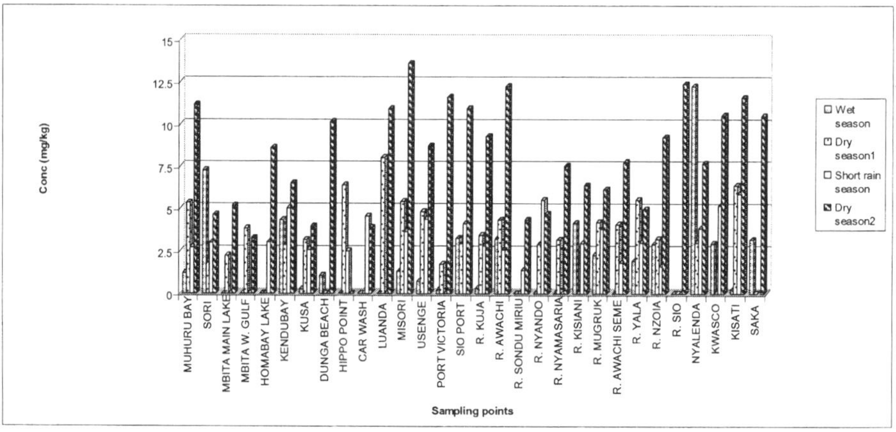

Figure 2 *Sediment bio-available phosphates*

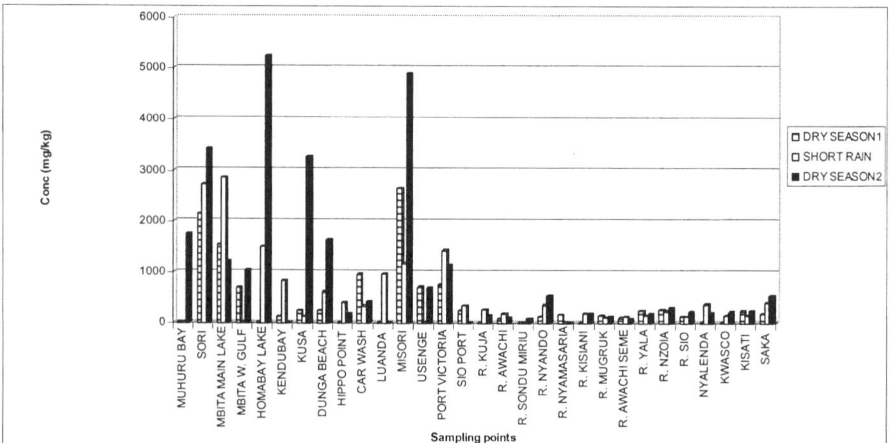

Figure 3 *Soil-available phosphates*

deposited in different ways. PCB poisoning can lead to liver damage, respiratory disorders, various neurological symptoms and reproduction disorders. Hence, it is of great importance to assess the source of PCBs that may be detrimental to the lake ecosystem and ultimately to man. Figure 4 shows the results of PCB 28 analysis along the Kavirondo Gulf (Kenyan side) of Lake Victoria.

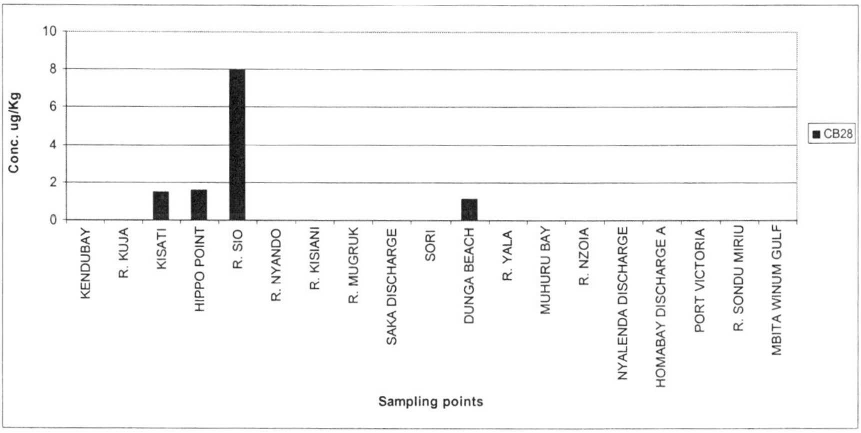

Figure 4 *Concentration of PCB 28 in sediments at different sampling stations in Kavirondo Gulf of Lake Victoria catchments*

High concentrations of PCB 28 were found at River Sio, Hippo Point, Kisati, and Dunga Beach. The curious part is the source of the high levels of PCBs in River Sio, since this river mainly drains the rural areas largely constituting agricultural parts of Western Kenya, with main industries being the cotton gins. Figure 5 shows the concentrations of PCB 52. The highest concentrations were found at the mouth of River Kuja and River Kisiat. River Kuja drains the Kisii highlands where mainly high agricultural activities take place while River Kisiat drains the industrial area of Kisumu Town. The third highest concentration was found at Hippo Point, which is close to the sewage discharge point of Kisumu Town. Low concentrations were observed at other sampling points.

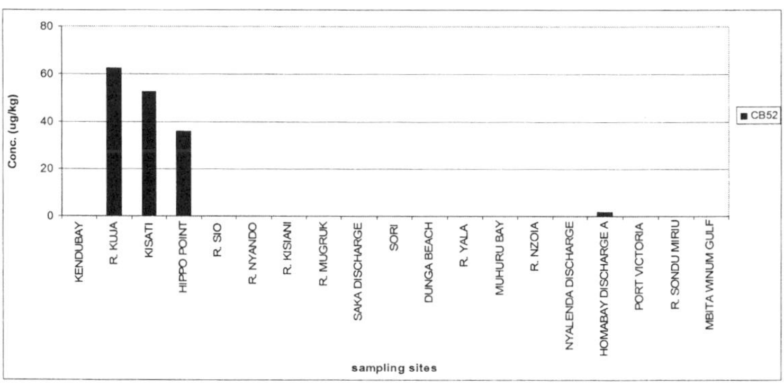

Figure 5 *Concentration of PCB 52 in sediments at different sampling stations of Kavirondo Gulf of Lake Victoria catchments*

PCB 101 was detected in most of the sediments from the drainage basin except for Saka and Nyalenda. The highest concentration of PCB 101 was detected in samples from River Sio, Muhuru Bay and Sori Bay. Figure 6 summarises the levels of PCB 101 detected at various points.

We also carried out analysis for PCB 105 at the shore sampling sites of the Kavirondo Gulf of Lake Victoria. Figure 7 gives the analysis results. The highest concentrations of PCB 105 was found in the sediments at the mouth of the river Sio, followed by Mbita point of Winam Gulf and Muhuru Bay. Generally, high concentration of PCB 105 is observed at the river mouths and at the Bays.

Further analysis of representative congeners of PCB included that of PCB 118. Once again, the highest concentration of PCB 118 was in the sediments of the mouth of the river Sio, followed by Mbita point of Winam Gulf and Muhuru Bay. Generally, high concentration of PCB 118 is observed at the river mouths and at the Bays. It is of interest to note that the sewage concentration pond at Nyalenda in Kisumu Town had very low PCB 118. Otherwise this compound was detected at most of the sampling sites (Figure 8).

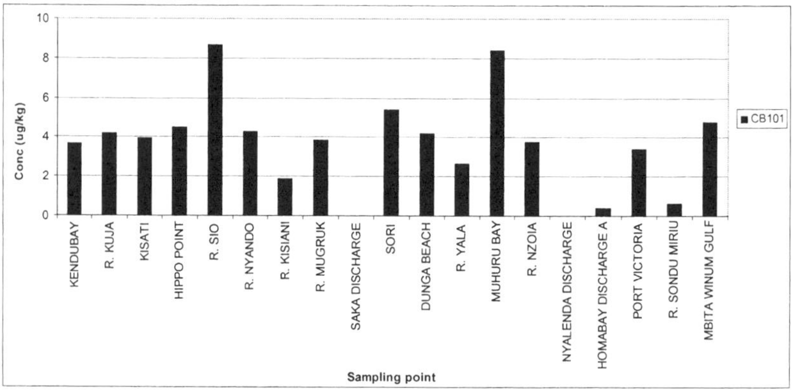

Figure 6 *Concentrations of PCB 101 in sediments at different sampling stations of Kavirondo Gulf of Lake Victoria catchments*

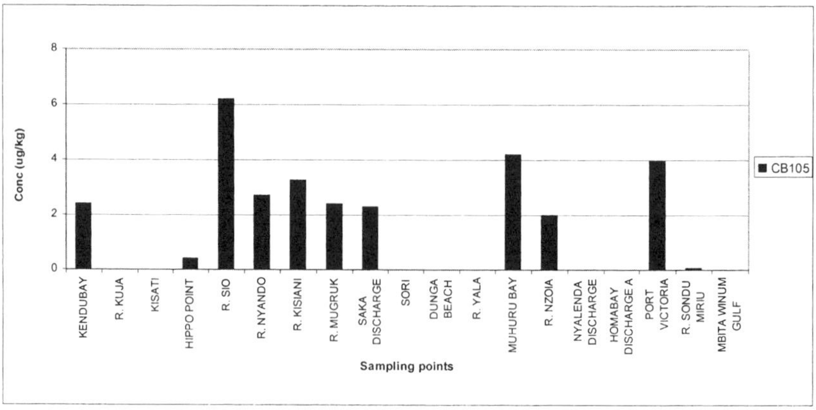

Figure 7 *Concentrations of PCB 105 in sediments at different sampling stations of Kavirondo Gulf of Lake Victoria catchments*

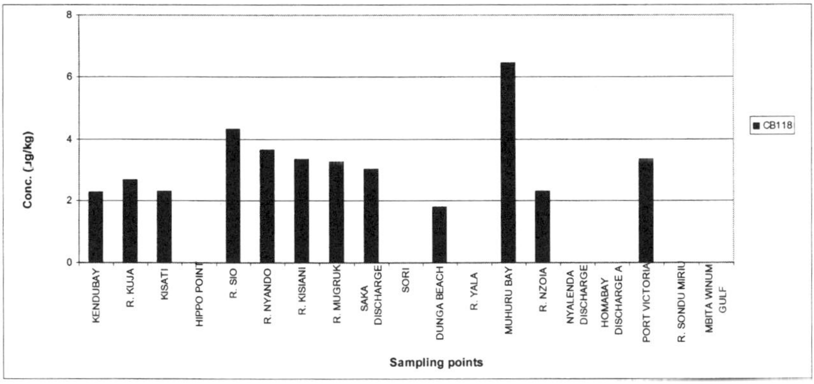

Figure 8 *Concentrations of PCB118 in sediments at different sampling stations of Kavirondo Gulf of Lake Victoria catchments*

Analysis for PCB 153 is given in Figure 9. Again it was found that the highest concentration of this compound was at River Sio mouth, followed by Muhuru Bay and Sori Beach. Again the sewage discharge points of Homa Bay and Nyalenda concentration effluent had low concentrations, indicating that the pollution by PCBs does not arise from sewage treatment points.

The last two PCBs analysed were PCB 156 and PCB 180 (Figures 10 and 11). The concentration of PCB 156 was highest in the sediments of the river Sio mouth followed by River Kisiani. On the other hand PCB 180 was highest in the sediments of Hippo Point, River Sio and Kisian River.

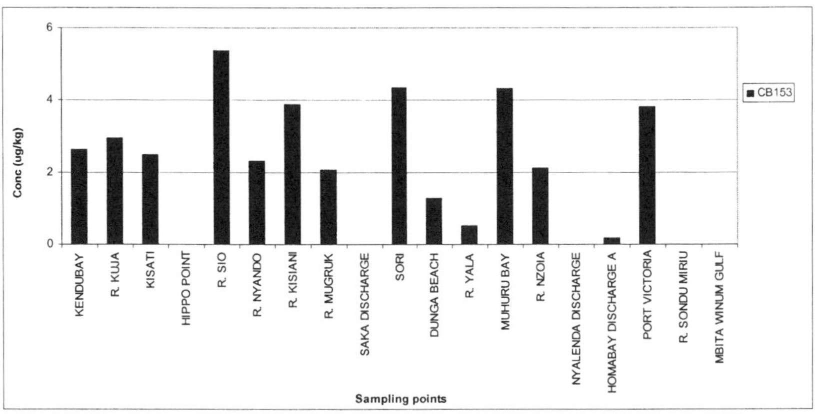

Figure 9 *Concentrations of PCB 153 in sediments at different sampling stations of Kavirondo Gulf of Lake Victoria catchments*

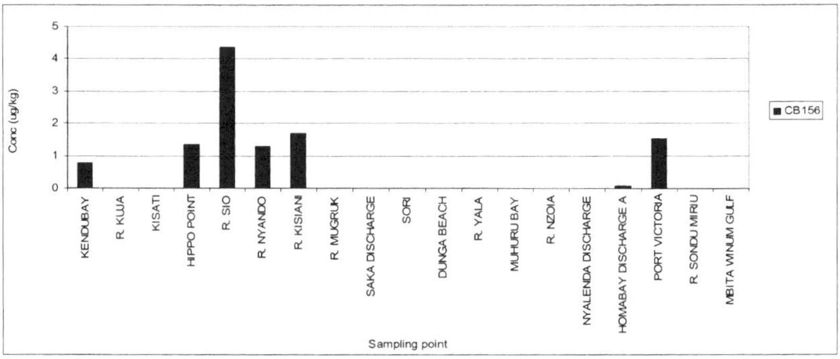

Figure 10 *Concentrations of PCB 156 in sediments at different sampling stations of Kavirondo Gulf of Lake Victoria catchments*

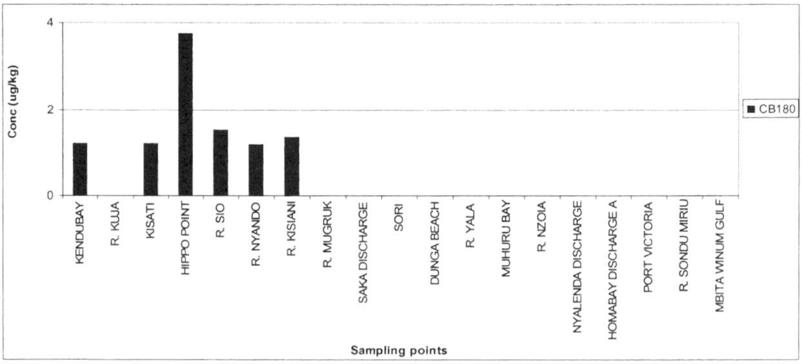

Figure 11 *Concentrations of PCB 180 in sediments at different sampling stations of Kavirondo Gulf of Lake Victoria catchments*

The observation of PCBs at almost all sampling points poses a challenge as to the source of the compounds. These are rural areas with very little electrification points. At this time it is speculative to attribute the source to any one activity except to say that there may be unauthorized dumping of PCBs in the catchments. Further studies are needed to pinpoint the point source of PCBs in the Lake Victoria waters and sediments. However, at present the data indicate that river mouths have higher concentrations of PCBs except for a few bays that are unique, that is Hippo Point, Sori and Port Victoria which are known for active commercial activities.

3.2.2 Pesticides in the Lake. Kenya heavily depends on the use of agrochemicals for the economic management of the agricultural production. Pesticides are used to control the myriad of insect pests, weeds, and disease pathogens that emerge due to the hot and humid tropical environmental conditions. The public health sector in Kenya also depends on the use of pesticides to control vector-borne diseases such as malaria, sleeping sickness, bilharziasis and fascioliasis. The pesticide spray programs aimed at controlling disease vectors such as mosquitoes, tsetse-flies and water snails succeeded to render Mwea Tabere settlement scheme, Kano plain and Lambwe Valley habitable using p,p'-DDT, dieldrin and endosulfan. Most organochlorine pesticides have been banned from agricultural use due to their high levels of persistence and toxicity to non-target organisms, and have been replaced by organophosphates, carbamates and pyrethroids. However, their residues are still in the environment.

Various studies in Kenya have reported organochlorine pesticides residues in a number of rivers and lake ecosystems at varying concentrations and frequencies. Getenga *et al.*[17] reported α-BHC, β-BHC, lindane, endosulfan, heptachlor, aldrin, heptachlor epoxide, dieldrin, endrin and methoxychlor residues in water samples collected from rivers draining the sugarcane fields and in soils from sugarcane fields that constitute the Lake Victoria basin. Of all the detected residue levels, lindane and α-BHC were at the highest concentration, with α-BHC showing an average concentration of 0.219±0.091 mg/l.

Mitema and Gitau[18] reported low levels of α-BHC, β-BHC, aldrin, dieldrin, lindane and p,p'-DDT in Nile perch from Lake Victoria. The p,p'-DDT and its metabolites formed the largest proportion of the organochlorine pesticide residues in the fish samples. The presence of these residues was attributed to the previous use of the pesticides in agriculture and aerial control of mosquitoes in the Lake Victoria region. Studies conducted in other parts of the country revealed similar trends of pesticide contaminations. Wandiga[19] reported residues of aldrin, α-endosulfan, dieldrin, endrin, DDT, DDE, DDD and lindane in seawater, seaweeds, sediments, and fish. They found residue levels ranging from 0.503-9.025 ng/g in sea water, 0.584-59.00 ng/g in sediments, and concentrations of 1011 ng/g and 418 ng/g of p,p'-DDT and p,p'-DDD, respectively, in fish.

Everaarts[20] detected the presence of PCBs and cyclic pesticides in benthic organisms from the Kenyan coast. They reported higher concentrations of PCB congeners and cyclic pesticides at the mouth of Sabaki River than the mouth of Tana River. The bivalve molluscs from the mouth of Sabaki River and Kiwaya Bay had the highest levels of PCBs. Residues of p,p'-DDE were detected in all the samples at levels ranging from 15 to 48 ng/g of lipid in both bivalve and gastropod molluscs.

Studies by Wandiga[22] on sediment samples reported residues of lindane, aldrin, p,p'-DDT, and p,p'-DDE in most of the samples analysed. They found that endosulphan, dieldrin, p,p'-DDD and endrin were either present in only low concentrations or were absent in most of the samples. They reported residue levels ranging from 0.584-59.00 ng/g of sediments.

The presence of pesticides at the top of the food chain has been reported in some studies and may be of critical concern. Pesticide residues have been detected in samples including cows' and human milk, and birds. Kituyi[21] found contamination of the cows' milk by chlorfenviphos residues in levels ranging between 0.52 and 3.90 mg/kg in dry season and 1.58 to 10.69 mg/kg during wet season. The same study showed that milk collected from plunge-dipped cows had higher concentrations of the pesticide residues than milk obtained from hand-sprayed cows.

In order to understand the behaviour of pesticides applied in the tropical environment, a number of simulation studies as well as field experiments have been carried out on different ecosystems in Kenya, and have been reviewed.[22]

In studies conducted in other countries, high levels of DDT and HCH isomers were reported in human serum samples in India.[23] p,p'-DDE, o,p'-DDT, p,p'-DDD, p,p'-DDT, (∑DDT=32.61±2.32 μg/l), as well as α-HCH, β-HCH, and γ-HCH (∑HCH=41.23±3.77 μg/l), were found, whereas HCB residues were 0.20±0.002 μg/l. Other studies[24] conducted in Canada reported medium and maximum whole blood levels of HCB as 0.11 and 0.34 ppb. Field studies in Uganda and Tanzania also revealed pesticide contamination.[25,26] Concentrations ranging from 0.01-0.03 mg/L of α-endosulfan, β-endosulfan and endosulfan sulphate were detected in fish from Tanzania.[27]

In general, the presence of a number of pesticide residues in the aquifer systems and in the food chain is causing great concern. This is due to the fact that many of the older synthetic pesticides have high ability to bio-concentrate and bio-accumulate in the food chain and thus cause long-term impact to the sustainability of the ecosystem. As a result, there is need to monitor the levels and extents of contamination of the environment by these compounds. This study was set up to investigate the influence of the catchments to the influx of organochlorine pesticide residues in the River Nzoia drainage system of Lake Victoria.

We describe here below the results of our study in terms of seasons. The use of pesticides is less intensive in the dry season than in the wet season.

3.2.2.1 Short rain season

Water samples

The samples collected during the short rain season showed the highest levels of pesticide residues, of which the residues of p,p'-DDT, HCH, dieldrin and heptachlor were the highest detected organochlorine pesticides (Figure 12).

The detected residue levels of p,p'-DDT ranged from below the detection limit (BDL) to 1.13 μg/l, while γ-HCH was between BDL to 0.32 μg/l. Dieldrin was detected in all the samples and its concentration ranged between 0.12 to 0.6 μg/l. This was slightly higher than that detected for aldrin, which ranged from 0.1 to 0.15 μg/l. The levels of endosulphan sulphate were between 0.02 and 0.4 μg/l, compared to those of alpha endosulphan that were between 0.01 and 0.09 μg/l, suggesting potential degradation process of the latter. The levels of heptachlor were between BDL and 0.59 μg/l, which was higher than those of heptachlor epoxide, which were found at concentrations between 0.11 and 0.13 μg/l. This was attributed to illegal use of this compound in the environment.

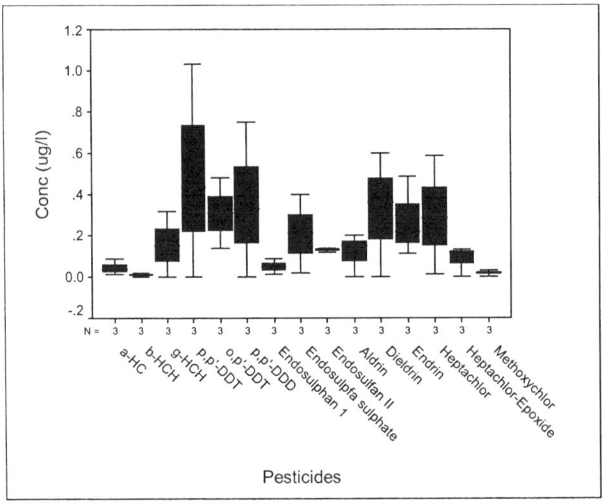

Figure 12 *Organochlorine pesticide residues in water during short rain season*

In general, the levels of p,p'-DDT, HCH and methoxychlor residues were much below the World Health Organization (WHO) maximum limits for the drinking water, whereas aldrin, dieldrin, heptachlor and heptachlor were above the WHO guidelines in some of the samples.[28] The levels of γ-HCH were greater than those of α-HCH, indicating current use of lindane in the catchment. However, the levels of dieldrin and endosulphan sulphate were much higher than those of α-endosulphan indicating that there was no recent application of these compounds. High levels of heptachlor in the water sample indicated that the compound might be in use by some farmers. However the use of heptachlor in the country was banned a long time ago due to its persistence and toxicity effects to non target species. Heptachlor in the environment breaks down to heptachlor epoxide. We could have expected higher levels of the epoxide compared to the heptachlor itself. Therefore, the high levels of heptachlor in the samples could imply that some farmers are illegally applying pesticides containing this compound.

Sediment samples

The analysis of the sediment samples indicated pesticide residues of magnitude over ten times higher than these detected in the water samples (Figure 13). This indicated that sediments accumulate more pesticide residues due to their high organic carbon content, which enhances sorption of the hydrophobic compounds to the sediment surfaces.

Dieldrin was found at higher concentrations than any other pesticides analysed in sediments collected during the short rain season, with concentrations ranging from 43.21 to 60.63 μg/kg. The level of aldrin in the sediments ranged from 9.14 to 9.68 μg/kg and was attributed to a degradation process, since aldrin in the environment breaks down to dieldrin.

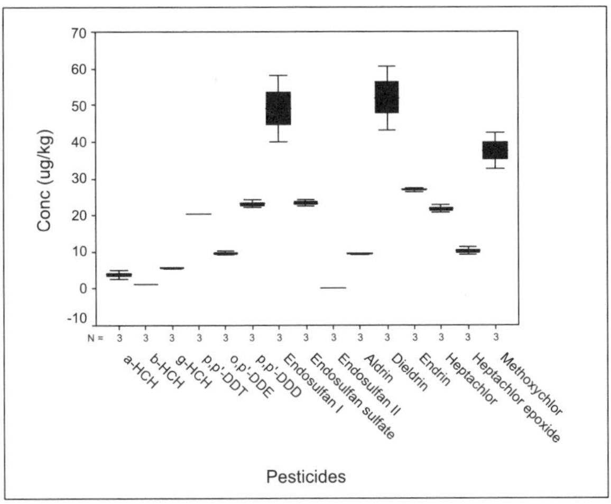

Figure 13 *Pesticide residues in sediments during short rain season*

The concentration of γ-HCH, 5.39-5.73 μg/kg, was higher than that of α-HCH, which ranged from 2.64 to 5.1 μg/kg. In comparison with the trend observed for HCH residues in the water samples, this indicated that HCH is still in use in the region. The residue levels of p,p'-DDD ranged between 22.03 and 24.25 μg/kg and were higher than those of p,p'-DDT (20.32-20.52 μg/kg), indicating that DDT was not in current use in the region.

In general, α-endosulphan and dieldrin were the highest residues detected in the sediment samples. The high residues of methoxychlor found in the sediment samples were attributed to adsorption processes which tend to retain hydrophobic compounds in the sediments and soil surfaces compared to the water column. In general, the levels of p,p'-DDD and dieldrin were higher than those of p,p'-DDT and aldrin, indicating no recent use of these pesticides in the region. The levels of γ-HCH detected in the sediments were higher than those of α-HCH and this indicated that HCH might be still in use by some farmers. However, the use of HCH in agriculture was banned in Kenya and thus there is need to monitor the sources of these residues in the environment.

3.2.2.2 Dry season

Water samples

The levels of pesticide residues detected in the water samples collected during the dry season were lower than those detected during the short rain season. This indicated that most of the organochlorine residues are swept from the agricultural land, where they were previously applied, by the storm water into the drainage systems as observed in the short rain season. Dieldrin, o,p'-DDE and heptachlor epoxide were detected in all the water samples collected during the dry season (Figure 14).

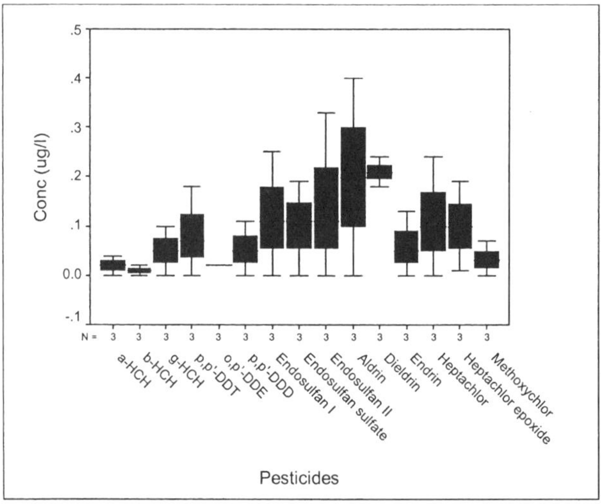

Figure 14 *Pesticide residues in water during dry season*

The levels of HCH, DDT, and methoxychlor were below the WHO guidelines for drinking water in all the samples analysed. However, some samples contained levels of aldrin, dieldrin, α-endosulphan and heptachlor above the recommended guidelines in some of the samples. The residue levels of dieldrin ranged between 0.18 and 0.24 μg/l, whereas those of aldrin were between BDL and 0.4 μg/l.

The presence of dieldrin in all the samples was attributed to the breakdown of aldrin that was previously applied in the region. This was confirmed by the dieldrin-aldrin ratio which was 1.05, indicating no recent use of aldrin in the region. The concentrations of p,p'-DDT ranged between BDL and 0.18 μg/l compared to those of p,p'-DDD which were between BDL and 0.11 μg/l. A comparison of levels of HCH isomers indicated that γ-HCH was between BDL and 0.1 μg/l, whereas the α-HCH had concentrations ranging between

BDL and 0.04 μg/l. In reality, γ-HCH breaks down to α-HCH, therefore the high level of the former indicates that lindane is still in use by some farmers in the region.

The residues of endosulphan were higher than those of endosulphan sulphate. This also indicated that endosulphan is in use in the region. The use of endosulphan is not banned in the country, and this may contribute to its high use by the farmers. The residues of heptachlor, ranging between BDL and 0.21 μg/l, were greater than those of heptachlor epoxide. However, heptachlor epoxide was detected in all the samples, and was attributed to possible degradation of heptachlor (Figure 14). In general, there is need for a monitoring program to establish whether there are cases of illegal use of banned compounds by the farmers in the region.

Sediment samples

The residues detected in the sediment samples collected during the dry season were higher than the levels in the water samples as shown in Figure 15. However, they were slightly lower than those detected in the sediment samples collected during the short rain season. This was attributed to degradation processes and transport by the moving water downstream.

The residues of dieldrin were the highest of all the organochlorine pesticides analysed, with concentrations ranging from 36.42 and 43.4 μg/kg. Compared with the values detected for aldrin, whose concentration ranged between 25.16 and 30.36 μg/kg, the high levels of dieldrin were attributed to degradation of the latter.

The levels of p,p'-DDD ranged between 21.16 and 23.35 μg/kg, compared to those of p,p'-DDT, which were between 18.91 and 19.6 μg/kg, also indicating possible degradation. The analysis of endosulphan in the sediments collected during the dry season showed that α-endosulphan levels ranged between 10.8 and 11.28 μg/kg, whereas endosulphan sulphate ranged between 9.97 and 10.67 μg/kg. The trend was similar to that detected in water samples, confirming that endosulphan is used in the region.

In general, the p,p'-DDD, dieldrin and heptachlor epoxide residues were higher than those of p,p'-DDT, aldrin, and heptachlor. This implied that there was no recent application of DDT, aldrin and heptachlor. However, the residues of γ-HCH and α-endosulphan were higher than those of α-HCH and endosulphan sulphate, respectively. This indicated that there might have been recent application of these compounds in the environment. The residue levels detected in sediments were in comparable range with those reported by Wandiga *et al.*[22] in sediment samples from the coastal part of Kenya.

3.2.2.3 Heavy rain season

Water samples

The residue levels of pesticides in the water samples collected during the heavy rain season were the lowest of all the three seasons (Figure 16). This was attributed to the fact that the samples in this season were collected five months after the short rain season, and the large volume of precipitation water associated with the heavy rain season could have increased the dilution factor of the pesticides and thus lowering the concentrations in the samples.

The levels of DDT, HCH and methoxychlor were below the WHO limit for all the samples analysed. However the levels of dieldrin, aldrin, α-endosulphan and heptachlor were still above the WHO limit in some of the samples.

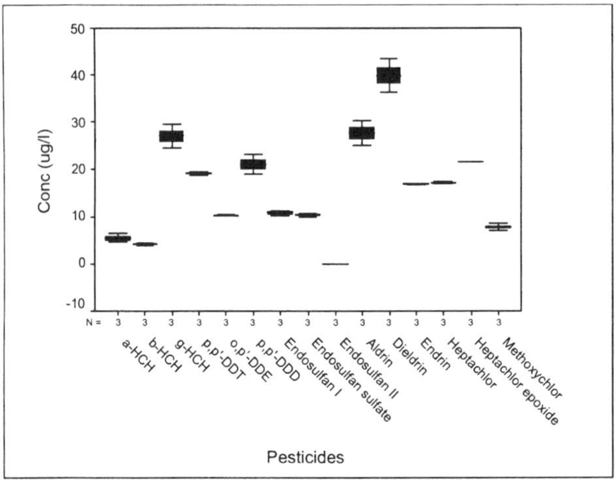

Figure 15 *Pesticide residues in sediments during dry season*

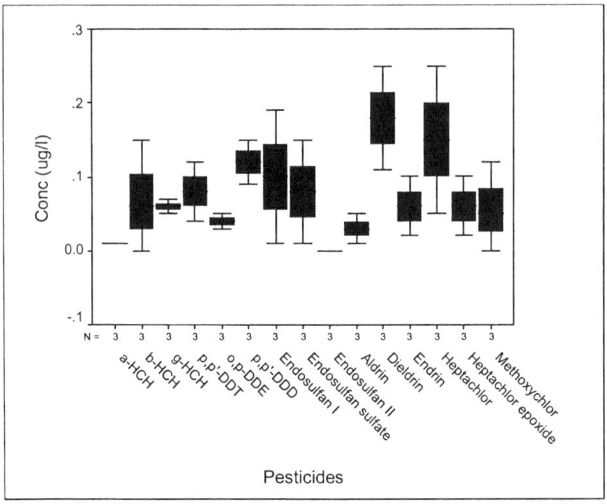

Figure 16 *Pesticide residues in water during heavy rain season*

The concentration of p,p'-DDT detected in samples collected during the heavy rain season ranged between 0.04 and 0.12 μg/l. The level of p,p'-DDD was higher, with concentrations ranging between 0.09 and 0.15 μg/l. The observed trend was attributed to the degradation of p,p'-DDT to p,p'-DDD. Generally, DDT breaks down in the environment to DDE and DDD. The final product depends on the environmental conditions, leading to aerobic or anaerobic degradation.

The analysis of dieldrin showed concentrations between 0.11 and 0.25 μg/l, whereas aldrin ranged between 0.01 and 0.19 μg/l. The concentration of α-endosulphan residues in the samples collected during the heavy rain season was between 0.01 and 0.15 μg/l whereas endosulphan sulphate was between 0.01 and 0.15 μg/l. In general, the analysis of the metabolites showed that the levels of p,p'-DDD, dieldrin were higher than those of p,p-

DDT and aldrin, indicating no recent use of these compounds. However, the levels of γ-HCH and α-endosulphan were still greater than those of a-HCH and endosulphan sulphate. This indicated that those pesticides could still be in use in the region, despite the fact that they were banned from agricultural use.

Sediment samples

The analysis of sediment samples collected during the heavy rain season indicated the lowest residue levels of all the seasons that were covered. The residues of p,p'-DDD were the highest of all the organochlorines analysed in the sediments collected during the heavy rain season. This was attributed to previous use of DDT, which breaks down to DDE and DDD, in the region. Similarly, the levels of dieldrin and endosulphan sulphate were detected in the sediments compared to their parent compounds, aldrin and α-endosulphan, indicating the previous use of these compounds (Figure 17).

Analysis of individual pesticide residues showed that p,p'-DDD was leading, with concentrations between BDL and 12.25 μg/kg, followed by o,p-DDE and dieldrin. The levels of dieldrin ranged between BDL and 09.21 μg/kg, compared to those of aldrin, whose concentration ranged between BDL and 7.86 μg/kg. The residue levels of γ-HCH were higher than those of α-HCH, with concentrations ranging between BDL and 8.86 μg/kg, compared to BDL and 2.03 μg/kg detected for α-HCH, which implied that lindane is still in use in the region.

The levels of α-endosulphan ranged between BDL and 4.03 μg/kg, whereas endosulphan sulphate ranged between BDL and 4.4 μg/kg. This was in comparable range. The residues of heptachlor were detected at concentrations higher than those detected for heptachlor epoxide. In general heptachlor degrades to heptachlor epoxide. Therefore the higher levels of heptachlor compared to the epoxide would indicate that probably some farmers could be applying the formulated compounds containing heptachlor compounds. This calls for consistent monitoring of these residues in the environment since the use of most of these products for agricultural purposes has been banned.

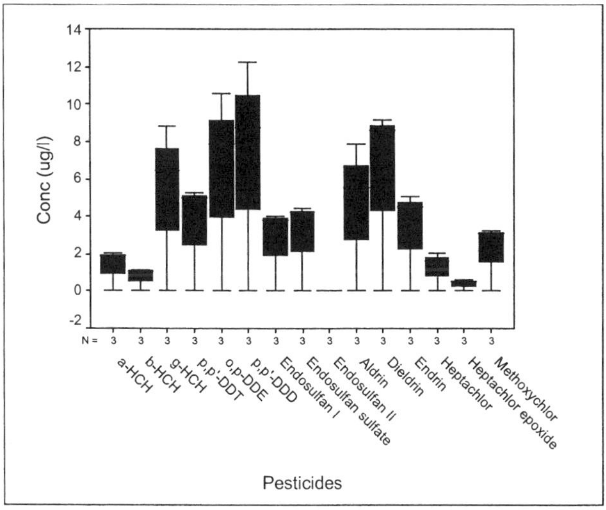

Figure 17 *Pesticide residues in sediments during heavy rain season*

The use of pesticides in Kenya dates back to the colonial era. The major areas of application include agricultural pest control and the public health control of mosquitoes, snails, and tsetse flies. However, with the rapid development of resistance of pests to the chemicals, coupled with environmental persistence, most of the organochlorines have been banned for agricultural use and restricted in the case of public health vector control. Nevertheless, there are detectable residue levels of these compounds in amounts that lead to concern for food chain magnification. These levels arise either from previous application or unscrupulous use through illegal means.

As a consequence, our research group has been involved in the assessment of the residue levels in both marine and freshwater ecosystems, as well as in the determination of persistence, toxic effects to fish and the transport of these compounds in different soils in the country. A comparative review of the pesticides, phosphates and polychlorinated biphenyls residues detected in water, sediments and biota from fresh water ecosystems and their effect on fish species has been presented. It gives comparative levels of a number of pesticides studied that include aldrin, dieldrin, α-endosulfan, endrin, DDT, DDE, DDD, PCBs and lindane for the period between 1998 and 2004. The previously detected residue levels in marine samples ranged from 0.503 to 9.025 μg/l in seawater, 0.584 to 59.0 μg /kg in sediments and BDL to 1011μg /kg in biota. The residue levels in freshwater ecosystem ranged between BDL and 0.44 μg/l in water, BDL and 65.48 μg/kg in sediments, BDL and 10.07 μg/kg in weeds, and BDL and 481.18 μg/kg in fish samples.

Other challenges facing water quality in Kenya include climate change and variability, industrial and municipal wastes, agricultural practices and deforestation. These have been highlighted to give the magnitude of the problem.

Acknowledgements

Vincent O. Madadi is highly acknowledged for the studies conducted for the PhD degree of the University of Nairobi, Department of Chemistry, in the author's laboratory.

References

1 M. Hulme, R.M. Doherty, T. Ngara, M.G. New and D. Lister, *Climate Research*, 2001, **17**, 145-168.

2 Intergovernmental Panel on Climate Change (IPCC), Fourth Assessment Report. Climate Change 2007: Synthesis Report. Summary for Policymakers. http://www.ipcc.ch/pdf/assessment-report/ar4/syr/ar4_syr_spm.pdf

3 D. Verschuren, T.C. Johnson, H.J. Kling, D.N. Edgington, P.R. Leavitt, E.T. Brown, M.R. Talbot and R.E. Hecky, *Proc. Royal Soc. London B*, 2002, **269**, 289.

4 R.E. Schulze, *AMBIO: A Journal of the Human Environment*, 2000, **29**, 12.

5 R.E. Hecky, *Proc. Int. Assoc. Theor. Appl. Limnol*., 1993, **25**, 39.

6 LVEMP (2001) Study on toxic chemicals/oil products spill contingency plan for Lake Victoria RFP#LVEMP/RCON/003 Volume V of the Toxic Chemical/ Oil Products Spill Contingency Plan "Waste Water Report" (2001).

7 LVEMP, (2002) Lake Victoria Environmental Management Project Phase, Water Quality and Ecosystem Management Component, Preliminary Findings of Studies Conducted on Lake Victoria.

8 R.A. Tamatamah, R.E. Hecky and H.C. Duthie, *Biogeochemistry*, 2005, **73**, 325.

9 P.A.G.M. Scheren, H.A. Zanting and A.M.C. Lemmens, *Journal of Environmental Management*, 2000, **58**, 235.

10 COWI Consulting Engineers (2002). Integrated water quality/limnology study for Lake Victoria Environmental Management Project, Part II. Technical Report.
11 Guidelines on the use of urine and faeces in crop production. http://www.ecosanres.org/pdf_files/ESR-factsheet-06.pdf
12 P.S. Hooda, A.C. Edwards, H.A. Anderson and A. Miller, A review of water quality concerns in livestock farming areas, *The Science of the Total Environment*, 2000, **250**, 143.
13 G. Morgan, Q. Xie and M. Devins, (2000) Small catchments –NMP Dripsey – water quality aspects. *In:* Tunney, H. (ed.) *Quantification of Phosphorus Loss from Soil to Water*, EPA, Wexford, pp. 24–37.
14 A.N. Sharply, B. Foy and P. Withers, *J. of Environmental Quality*, 2000, **29**, 1.
15 H. Tunney, B. Copulter, K. Daly, I. Kurz, C. Coxon, D. Jeffrey, P. Mills, G. Kiely and G. Morgan (2000), *Quantification of phosphorus loss from soil to water*, EPA, Wexford.
16 S.O. Wandiga and V.O. Madadi (2009). Chapter 3. Water Quality Issues in Eastern Africa. In. S. Ahuja, ed. Handbook on water purity and quality. Academic Press Inc. NY., p39-65. In Press.
17 Z.M. Getenga, F.O. Kengara and S.O. Wandiga, *Bull. Environ. Contam. Toxicol.*, 2004b, **72**, 335.
18 E.S. Mitema and F.K. Gitau, *African J. Ecol.*, 1990, **28**, 234.
19 S.O. Wandiga, P.O. Yugi, M.W. Barasa, I.O. Jumba and J.O. Lalah, *Environmental Toxicology*, 2002, **23**, 1235.
20 J.M. Everaarts, E.M. vanWeerlee, C.V. Fischer and M.T.J. Hillebrand, (1997) Polychlorinated biphenyls and cyclic pesticides in sediments and micro invertebrates from the coastal regions of different climatological zones. In: *Environmental behaviour of crop protection chemicals, Proc. Int. Symp. on use of nuclear related techniques for studying environmental behaviour of crop protection chemicals.* 1-5 July 1996, IAEA-SM-343/45, Vienna: IAEA/FAO, pp. 407-431.
21 S.O. Wandiga, D.M.K. Ongeri, L. Mbuvi, J.O. Lalah and I.O. Jumba, *Environmental Toxicology*, 2002, **23**, 1285; Use and distribution of organochlorine pesticides in Africa (2001). S.O. Wandiga. *Pure and Applied Chemistry* **73 (7)** 1147 - 1155.
22 E. Kituyi, I.O. Jumba and S.O. Wandiga, *Bull. Environ. Contam. Toxicol.*, 1997, **58**, 969.
23 V.K. Bhatnagar, R. Kashyap, S.S.A. Zaidi, P.K. Kulkarni and H.N. Saiyed, *Indian Bull. Environ. Contam. Toxicol.*, 2004, 72, 261.
24 J. Mes, *Bull. Environ. Contam. Toxicol.*, 1992, **48**, 815.
25 S.B. Mbabazi, (1998). Levels of lindane in soil, plants, and selected soil fauna in a sprayed in maize agroecosystem. M.Sc. Thesis, Makerere University, Kampala, Uganda.
26 H. Mwevura, O.C. Othman and G.L. Mhehe, *Tanz. J. Sci.*, 2002, **28**, 117.
27 L. Henry and M.A. Kishimba, (2000) Levels of pesticide residues in southern Lake Victoria and its basin. Lake Victoria 2000: A New Beginning; *International Conference*, Jinja.
28 WHO, (2000) World Health Organization Guidelines and International Standards for Drinking-Water Quality. ACQWS paper No. 3.

FOUNDATION FOR ANALYTICAL SCIENCE AND TECHNOLOGY IN AFRICA AND ITS ROLE IN THE PRESERVATION OF ENDANGERED SPECIES

S. Lancaster,[1] N. Richards[2] and A. Gachanja[3]

[1]Foundation for Analytical Science & Technology in Africa, 49 Swanland Road, Hessle, UK, HU13 0NN.
[2]Department of Forensic Science, Anglia Ruskin University, East Road, Cambridge, UK CB1 1PT.
[3]Chemistry Department, Jomo Kenyatta University of Agriculture & Technology, P.O. Box 62000, 00200 Nairobi, Kenya.

1 INTRODUCTION TO FOUNDATION FOR ANALYTICAL SCIENCE AND TECHNOLOGY IN AFRICA

Foundation for Analytical Science and Technology in Africa (FASTA) is a registered charity comprised of industrial and academic scientists from the analytical sector in the UK. It was founded by Steve Lancaster from BP and Barrie Nixon from Mass Spec UK on September 20, 2006, to support scientific education, analytical research and the preservation of the environment in Africa via capacity-building and technology transfer. Several organisations, including: The Royal Society of Chemistry, The British Mass Spectrometry Society, Mass Spec UK, Pfizer, BP and Perkin-Elmer, have been very supportive by providing generous grants and equipment donations. This support has been critical in allowing FASTA to set up the first GC-MS laboratory at Jomo Kenyatta University of Agriculture & Technology in Nairobi.

The aims and objectives of FASTA are to:

- facilitate research and teaching into chemical systems and processes;
- facilitate research and teaching of environmental processes;
- enable / facilitate the provision of environmental monitoring services at the local and national level;
- promote and encourage food analysis in order to facilitate the import and export of foodstuffs.

To achieve these aims and to ensure that they are sustainable, FASTA aims to create a critical mass of local scientists with expertise in advanced analytical technology which can react rapidly to the demand for high quality analysis in the preservation of the environment. To this end, FASTA is working closely with the Royal Society of Chemistry and the Pan Africa Chemistry Network (PACN).

Over the last two years, FASTA have installed and currently maintain two GC-MS systems in the Chemistry Department at Jomo Kenyatta University of Agriculture and Technology (JKUAT) in Nairobi. These instruments are currently being used for post graduate research and analysis of environmental samples at JKUAT. FASTA has also introduced thermal desorption equipment to be interfaced with one GC-MS. This will facilitate detailed atmospheric research and monitoring for a range of atmospheric pollutants. Through this work, JKUAT is now recognised by the Kenyan government as a centre of excellence in environmental research and analytical chemistry.

FASTA scientists are collaborating with Professor Anthony Gachanja of JKUAT and continually widening the scope of the laboratory. The goal is for JKUAT, under the guidance of Professor Gachanja and the directors of FASTA, to become a hub for analytical science in East Africa, consisting of technical excellence and high-level analytical equipment which can be used for analysis and training throughout the region.[1-3]

1.1 FASTA and Biodiversity

The attrition of global biodiversity is a current topical issue and many worrying statistics outlining loss of habitat, species extinction and climate change are published in the press on an almost daily basis. Of great cultural, economic and ecological concern is the dramatic decline in vulture populations in Asia, as well as in the decline of large predators and scavenging species in Africa.

JKUAT and FASTA are now collaborating with Ms Ngaio Richards, a researcher and conservationist currently completing her doctoral work in the area of wildlife forensics. This work is resulting in the development of forensic methodology for the detection of veterinary drugs in livestock animals and vultures in Africa, following the well documented mass mortality of vultures observed in India. She is also liaising with a number of collaborators in East Africa, including Wildlife Direct, the National Museums of Kenya and Maseno University, on the broader and more immediate issue of human-wildlife conflict and wildlife (and human) poisoning in agricultural areas. FASTA and JKUAT are now collaborating with Ms. N. Richards so that the following may be achieved:

- develop a strategic sampling network to monitor and record incidences of wildlife mortality (with emphasis on vultures, initially) throughout East Africa;
- provide workshops and training on fundamental and current analytical and forensics methodology to laboratory, field and research personnel within the network;
- offer community workshops to farmers, veterinarians and local agri-businesses on the safe handling (for people and wildlife) and use of pesticides and other potentially dangerous compounds;
- encourage and support community endeavours to minimise human-wildlife conflict and reduce harmful impacts from pesticide use.

2 THE VULTURE CONSERVATION PROJECT

Ongoing exposure to the nonsteroidal anti-inflammatory drug (NSAID) diclofenac is now considered to be the sole or primary cause of the virtual extinction of three species of *Gyps* vultures on the Indian subcontinent.[4] Once among the most abundant species on Earth, these vultures rapidly and thoroughly scavenged the millions of livestock carcasses that were left out for them each year. The livestock, used as working animals, had been treated with diclofenac to treat pain and inflammation from lacerations and ailments related to old age. Wild dogs and rats have moved in to replace the vultures, but they are far less efficient as scavengers and actively pose a number of health risks to residents and to domestic and livestock animals. Because carcasses are now only partially consumed, people who made their living retrieving bones for fertiliser have to find another means of subsistence. Sadly, a funerary rite that involved offering human bodies to vultures to prevent the contamination of the environment can no longer be practiced.[5]

When certain species of vultures are exposed to diclofenac they suffer fatal kidney failure, identified by the presence of visceral gout.[6] Veterinary use of diclofenac is now being phased out on the Indian subcontinent, and the drug meloxicam has been identified as a safer and suitable alternative.[7] Interim actions include the capture of young wild birds and surviving adults for safekeeping and captive breeding programmes. The safety of other NSAIDs registered for veterinary use to vulture and other wildlife species remains to be assessed. Discussions are underway with the pharmaceutical industry on adopting rigorous safety testing and more appropriate risk assessments of these drugs, as well as potential new drugs that are in development.

Often maligned for their ungainly appearance and demeanour, vultures worldwide face various threats: human persecution, loss of habitat and prey base, secondary poisoning and exposure to contaminants. The situation on the Indian subcontinent challenged the often aesthetically-based notion of species value, demonstrating that these birds play an integral role in maintaining human, animal and ecosystem health. It has also generated an increasing interest in whether or not exposure to NSAIDs is playing a role in observed population declines elsewhere.

Diclofenac is presently analysed for in tissues. However, carcasses are not always found intact, and tissues may degrade unless retrieved quickly. To our knowledge, no methods have been developed for other NSAIDs. To address this, N. Richards is developing a multi-screening procedure to detect diclofenac and other NSAIDs of concern in animal hair, feathers and talons, via GCMS. It is known that drugs incorporated into human hair and nails are effectively locked in, and these structures are very resistant to environmental deterioration.[8-10] NSAID detectability is currently being investigated in feathers and bone. The objective of this research is to provide an alternate diagnostic and monitoring tool, to complement conventional tissue analysis and increase the number of carcasses that can be submitted and screened for exposure to NSAIDs.

2.1 Other concerns: the African perspective

Diclofenac is now registered for veterinary use and is readily available in some parts of Africa. Vigilance will be required on the part of conservationists and researchers to ensure that diclofenac and other NSAIDs used in Africa are not further contributing to the declines. Exposure to even a fraction of the treated carcasses that were available in India, for example at vulture 'restaurants' or feeding stations, could spell disaster for Africa's already imperilled vultures.

In Kenya, an onslaught of wildlife mortality has recently received media attention. Lions, hyenas, jackals, vultures and other birds of prey have perished by consuming carcasses baited with poison. Indiscriminate use of poisons and pesticides, particularly in response to human-wildlife conflict, are taking their toll on Kenya's charismatic wildlife, which also has serious repercussions for the tourist industry. The carbamate insecticide carbofuran is suspected in many of the poisonings.[11] There is also concern for human health as the insecticide may be improperly used for agricultural applications as well as for hunting.

Ultimately, these situations have exposed the dearth of baseline data that exists on human and wildlife poisonings and have shown the need to establish systematic and long-term monitoring of human and wildlife health in Kenya and throughout Africa.

2.2 Opportunity for conservation, collaboration and capacity-building

A pilot project is being set up with Professor Anthony Gachanja and the JKUAT laboratory to implement the NSAID detection methodology as part of specialised analysis of vulture samples, to be offered under the centre's environmental analytical services. Because they mask the signs of pain and inflammation, the use of these drugs is prohibited or closely monitored in the horseracing industry. Thus, the detection methodology will also be adapted for analysis of equine doping so that the lab can offer low cost analysis of wildlife samples to conservation bodies and other stakeholders with minimal financial resources while meeting its own costs. The JKUAT laboratory will liaise with relevant laboratories and researchers to remain apprised of developments in the conservation and analytical community. Additionally, FASTA is now involved with the Royal Society of Chemistry's initiative, the Pan Africa Chemistry Network (PACN), which is developing links between chemists across the entire African continent and facilitates collaboration with UK scientists. At the same time, there is a movement underway to comprehensively screen human and animal samples for evidence of exposure to a number of compounds of concern.

As scientists with a passion for chemistry and a dedication to the preservation of the environment, the directors of FASTA have made an enduring commitment to provide participating laboratories with the facilities they require to establish themselves and become self sustaining into the future. We offer our support and good wishes to the emerging analytical and research / conservation talents in Africa and hope to be involved in this exciting process of collaboration, capacity-building and conservation for many years to come.

References

1 S. Lancaster, *Mass Matters*, 2006, **51,** 11.
2 S. Crossland, *Mass Matters*, 2006, **52**, 6.
3 N. Richards and S. Lancaster, *Mass Matters*, 2007, **54**, 6.
4 D.J. Pain, C.G.R. Bowden, A.A. Cunningham, R. Cuthbert, D. Das, M. Gilbert, R.D. Jakati, Y. Jhala, A.A. Khan, V. Naidoo, J.L. Oaks, J. Parry-Jones, V. Prakash, A. Rahmani, S.P. Ranade, H.S. Baral, K.R. Senacha, S. Saravanan, N. Shah, G. Swan, D. Swarup, M.A. Taggart, R.T. Watson, M.Z. Virani, K. Wolter and R.E. Green, *Bird Conservation International*, 2008, **18**, S30-S48.
5 D.J. Pain, A.A. Cunningham, P.F. Donald, J.W. Duckworth, D.C. Houston, T. Katzner, J. Parry-Jones, C. Poole, V. Prakash, P. Round and R. Timmins, *Conservation Biology*, 2003, **17**, 661.
6 J.L. Oaks, M. Gilbert, M.Z. Virani, R.T. Watson, C.U. Meteyer, B.A. Rideout, H.L Shivaprasad, S. Ahmed, M.J.I. Chaudhry, M. Arshad, S. Mahmood, A. Ali and A.A. Khan, *Nature*, 2004, **427**, 630.
7 G. Swann, V. Naidoo, R. Cuthbert, R.E. Green, D.J. Pain, D. Swarup, V. Prakash, M. Taggart, L. Bekker, D. Das, J. Diekmann, M. Diekmann, E. Killian, A. Meharg, R.C. Patra, M. Saini & K. Wolter, *PLoS Biology*, 2005, **4**, 395.

8 V.A. Boumba, K.S. Ziavrou and T. Vougiouklakis, *Internat. J. Toxicology*, 2006, **25**, 143.
9 D.A. Engelhart and A.J. Jenkins, *J. Analytical Toxicology*, 2002, **26**, 489.
10 Y. Nakahara, *Journal of Chromatography B*, 1999, **733**, 161.
11 C. O'Driscoll, *Chemistry & Industry*, January 28, 2008, 5.

Jua Kali METALLURGICAL INDUSTRIES AS POINT SOURCES OF THE HEAVY METALS LEAD AND CADMIUM TO AQUATIC SYSTEMS IN KENYAN URBAN AREAS

S.M. Mule[1] and C.M. Nguta[2]

[1]Department of Chemistry, University of Eastern Africa, Baraton, Eldoret, Kenya.
Email: mules@ueab.ac.ke
[2]Chemistry Department, Egerton University, Njoro, Kenya.
Email: mweunguta@yahoo.com

1 INTRODUCTION

Jua Kali are Kiswahili words, which mean *hot sun*. They are commonly used to refer to small-scale enterprises operating on street pavements, open-air markets, undeveloped plots or market stalls.[1] They manufacture goods like metal boxes, pots, bicycle parts and jikos (cooking stoves) at affordable prices by low-earning individuals. Since 1972, the sector has grown tremendously, providing employment opportunities to millions of Kenyans.[2-4]

Eradication of poverty has been a major issue in Kenya. In the Sessional Paper No. 1 of 1986, in recognition of the contribution of *Jua Kali* enterprises, the Government projected that 6 million new jobs would have been created by the year 2000, of which 3.4 million would have been generated by the informal sector.[5,6] This led to the recognition of the contributions of *Jua Kali* enterprises, which had been neglected since 1972. In the Development Plan of Kenya, employment activities in the *Jua Kali* sector have been noted to be more than those in the formal wage sector in industry and commerce. In 1989, the government created an enabling environment for the *Jua Kali* sector by removing tedious administrative procedures required to obtain licenses and to ensure that large-scale firms did not interfere with the sector, and this has led to the rapid growth of these enterprises.[7,8]

The advent of *Jua Kali* enterprises in both big and small towns in Kenya is an encouraging phenomenon, when seen from the perspective of economic and social development of the nation. Such development prospects, however, could be threatening and may lead towards horrific environmental disasters unless emissions and wastes are monitored and disposed of at recommended safe sites.[5,9] Heavy metal pollution from these sites is of particular concern. Lead and cadmium are amongst the most toxic of these heavy metals, leading, for example, to interference with the function of the kidneys. Bone tissue is affected too, as cadmium disturbs the metabolism of calcium, and lead affects the central nervous system. In Kenya, much work has been done on the heavy metal content of soil and water, and of plant, fish, waterbuck and bird tissues.[10,11]

The manufacturing industries have been cited as being responsible for large emission loads of lead and cadmium. The manufacture of metallurgical products results in many and varied emissions, depending on the products being manufactured and the processes and chemicals used. Emissions to water include heavy metals (e.g. cadmium, mercury), organic micropollutants, suspended solids and organic matter. Manufacture processing also brings the risk of fire, explosions and spills, all of which may be a source of pollution of water. There is also a need to dispose of contaminated sludge from waste treatment plants.

Pure cadmium is a soft, silver-white metal found naturally in small quantities in air, water and soil. It does not have a definite taste or odor, and is not mined, but is a by-product of the smelting of other metals such as zinc, lead and copper. It is used in nickel-cadmium rechargeable batteries and for metal plating. It is also used in some paints, plastics and metal solders. Some metal containers, such as ice cube trays, pitchers, bowls or ceramic ware, can contain small amounts of cadmium. The main industries that use cadmium are the metal-smelting, electronics, nuclear power and paint pigment production industries, and other metal-working and refining companies.[6,7,9]

Almost all of the lead in soil comes from chips of lead-based paint flaking from homes, factory pollution, and the use of leaded gasoline. Over time, lead builds up in soil. The levels of lead in soil are usually higher in cities, near roadways and industries that use lead, and next to homes where crumbling lead paint has fallen into the soil. Soil naturally has small amounts of lead in it, but at levels of less than 50 mg/kg. Levels of greater than 500 mg/kg of lead have been reported in contaminated industrial soils.[5,7,10] Cadmium is released into the soil by metal processing industries, battery and paint manufacturing, and waste hauling and disposal activities. Aerial deposition of lead and cadmium from soot has been cited as another source of such metals in soil and water bodies.

The metals have the tendency to form compounds of low solubility with the major divalent cations (Pb^{2+}, Cd^{2+}) being found in natural water. Hydroxide, carbonate, sulfide, and, more rarely, sulfate may act as solubility controls in precipitating metal ions from water. A significant fraction of lead and, to a greater extent, cadmium carried by river water is expected to be in an undissolved form. This can consist of colloidal particles or larger undissolved particles of lead carbonate, lead oxide, lead hydroxide, or other lead compounds incorporated in other components of surface particulate matter from runoff. The ratio of lead in suspended solids to lead in dissolved form has been found to vary from 4:1 in rural streams to 27:1 in urban streams. The US Environmental Protection Agency (USEPA) has reported Maximum Contaminant Levels in water that are permissible to be 0.005 mg/L for cadmium and 0.015 mg/L of lead.[2,3]

However, no published literature is available for heavy metal contents contributed by *Jua Kali* enterprises to the environment. The results of this study will help the National Environmental and Management Authority (NEMA) to set control measures to reduce pollution due to heavy metals released from such sites to water bodies. The mushrooming of *Jua Kali* metallurgical enterprises in major towns in Kenya is of a major concern as they do not observe or adhere to any strict control in their discharges through effluents, emissions (atmospheric) and solid waste disposal with respect to these metals.[12,13] The purpose of this paper is to provide evidence that *Jua Kali* enterprises are a point source of heavy metals to the environment and therefore to recommend that mitigation measures should be put in place to control such pollution.

2 EXPERIMENTAL METHODS

2.1 Sampling Procedure

Three *Jua Kali* sites were selected from three different locations in Kenya, namely Kamukunji *Jua Kali* site in Nairobi city, Bondeni *Jua Kali* site in Nakuru town, and a Garage shed in Naivasha town. In all the *Jua Kali* sites visited, their main activities included the use of metal scraps to manufacture different goods, battery washing and charging, painting of metal bars and cars, metal foundry, and the burning of car tyres to get wires. It was also observed that in all sites the artisans were operating in shades and there was no proper way of disposing off wastes from such shades. Timboroa Forest was chosen as a control site in the study. Samples of soil, sediment and water were collected from all the sites, and were labelled, stored and transported to the laboratory for analysis.

Five grams of finely ground soil were weighed in a 250 ml Erlenmeyer flask to which was added 4 ml concentrated HCl and 15 ml concentrated HNO_3 sequentially. The same procedure was followed for the sediment and water samples. All digested samples were covered with convex glasses and left to stand overnight. The samples were digested at a temperature of 100°C for two hours and all mixtures were left to cool and then filtered. The filtrates were kept for elemental analysis using an Atomic Absorption Spectrometer (AAS) and all results were statistically analyzed by ANOVA tests. *In-situ* determination of water pH, Total Dissolved Solids (TDSs) and conductivity were measured using a SensIon 157 pH/TDS/Conductivity meter.

2.2 Case Study 1: Kamukunji *Jua Kali* Site

Kamukunji *Jua Kali* site is in Nairobi City. It is located several minutes' drive from the city center towards City stadium. Kamukunji *Jua Kali* shades are a hub of activities seven days a week. Noise from hammering of metals is adequate to draw the attention of any passer-by. Many of the activities undertaken by the artisans involve various metalworking activities, including in the manufacture of metal travelling boxes, braziers, pots, and bicycle parts. The artisans are crammed in small areas and each group is focused on its own activity. Oil and paint spillage is common on the site. Residential areas are close to the site and children are left to roam about in the site. Different types of fowls, such as chickens, were also noticed. Running waste-water passes through the site and into Nairobi River.

2.3 Case Study 2: Bondeni *Jua Kali* Site

Bondeni *Jua Kali* site is in Nakuru Town. It is located near the Bondeni Police Station along the Langa Langa road. The activities carried out range from metal working, battery charging, the welding of different kinds and types of products, the painting of different types of metal products, and, to a lesser extent, car repairs. Water runoff from the site is towards the famous Lake Nakuru.

2.4 Case Study 3: Naivasha Garage Shades

Not far from Naivasha town is a Motor Garage. Activities commonly found there range from painting, vehicle body building to mechanical repair. This largely enclosed place is cluttered with vehicles, which are being painted or repaired. The area is littered with discarded motor vehicle parts, and there is oil and paint spillage and a lot of noise from the various vehicles being repaired. Clogged water systems were evident. Spraying and painting vehicles are the major activities carried out by the *Jua Kali* workers, using motor-driven spray machines, the workers are engulfed by spray dust. The water draining from the Naivasha Garage runs into Lake Naivasha.

3 RESULTS AND DISCUSSION

The mean concentrations of lead in soil and sediment samples (Table 1) range between 59.1 mg/kg and 763.5 mg/kg. These values are well above those recorded for samples from the control site, which were 0.09mg/kg, and are within the values previously recorded for the levels of lead in contaminated soils and sediments.[3,4] Levels of lead in the water samples were between 0.983mg/L and 2.01mg/L, much higher than the levels of 0.006mg/L in the control water samples, and the USEPA maximum acceptable levels for standard water samples of 0.015mg/L. Similar values have been reported previously in contaminated water samples.[2,3,12]

These high values of lead in water at the *Jua Kali* sites is an indicator of contamination of the human beings and aquatic animals who use water from the water bodies for both domestic and agricultural purposes. Lead particles deposited in the soil from flaking lead paint, from incinerators, and from motor vehicles that use leaded gasoline, as well as water runoff and waste disposal from such *Jua Kali* sites, all contribute to lead to the environment.

Cadmium contributed by activities carried out by *Jua Kali* sites, such as battery washing, the burning of various kinds of metals, as well as paint spraying and welding, are contributing high levels of cadmium to the environment. The mean values of cadmium contributed to soil and sediment samples (Table 1) in selected *Jua Kali* sites range from 21.4mg/kg to 325.2mg/kg. The mean cadmium value from the control site was <0.03mg/kg. There was a high statistical correlation between the mean concentration values of lead and cadmium ($r = 0.9997$) in the *Jua Kali* metallurgical industries. Mean cadmium values in water samples ranged from 0.18mg/L and 0.32mg/L, and cadmium was not detected in water samples from control site. Reported values in the literature show that cadmium in water should be below 0.005mg/L.[3,5] Our results show that *Jua Kali* enterprises are a major source of cadmium to the environment.

Table 1 *Comparison of mean lead and cadmium levels in soil, sediment and water samples from* Jua Kali *sites and the control site (ND = not detected; n =35)*

Lead	Soil (mg/kg)	Sediment (mg/kg)	Water (mg/L)
Kamukunji	763.5 ± 6.6	87.4 ± 5.9	2.01 ± 0.04
Naivasha	488.3 ± 6.4	59.1 ± 4.4	0.983 ± 0.006
Bondeni	632.8 ± 8.9	68.2 ± 5.1	1.48 ± 0.08
Control	0.09 ± 6.4	ND	< 0.006
Cadmium			
Kamukunji	325.2 ± 3.5	21.4 ± 4.7	0.32 ± 0.05
Naivasha	184.9 ± 6.1	24.7 ± 3.1	0.18 ± 0.09
Bondeni	287.7 ± 9.9	30.1 ± 5.1	0.44 ± 0.07
Control	< 0.03	< 0.01	ND

Table 2 Mean *values of water pH, Total Dissolved Solids (TDS) and conductivity parameters* (n = 40)

	pH	TDS (mg/L)	Conductivity (μS/cm)
Kamukunji	5.38 ± 0.02	1980 ± 1.8	1057 ± 5.3
Naivasha	6.41 ± 0.09	2001 ± 3.4	997 ± 3.1
Bondeni	6.78 ± 0.04	1764 ± 4.6	852 ± 3.4
Control	7.02 ± 0.11	31 ± 1.1	15.4 ± 3.4
EPA Values	6.5-8.5	1500 max.	500-800

The pH levels of waste water from the *Jua Kali* sites are shown in Table 2. For two of the sites, the values are more acidic than those recommended by the EPA, while the values from the Bondeni and the control sites are within the recommended limits. Discharge of acid from battery charging sites can lead to low pH values.

The total dissolved solids is the total amount of dissolved matter in the water, e.g. calcium, magnesium, chlorides and sulfates. The values reported for the water in the above sites are higher than the values recommended by the EPA for drinking water, an indication that there are high amounts of dissolved substances. The values for TDS reported in Table 2 above are within the values for water classified as brackish (brackish water has TDS value >1500mg/L). Brackish water is hostile to the growth of most terrestrial plant species, and without appropriate management it is damaging to the environment, so its impact should be assessed. In addition, high TDS values cause scale build-up in the pipes, valves and filters of water systems, thereby reducing their performance and adding to system maintenance costs.

Values for water conductivity provide information about its chemical composition. The conductivity of water is directly linked to the concentration of ions within it and their mobility. The high concentrations of lead and cadmium in waste water displayed in Table 1 are an indication that there is a high concentration of ions in such waste water, leading to high conductivity.

4 CONCLUSIONS

The high concentrations of cadmium and lead in the environment contributed by *Jua Kali* sites cannot be ignored. Poor waste disposal practices, together with the lack of legislation, or the lack of implementation of legislation, at a time when *Jua Kali* sites are mushrooming, has contributed to the presence of these heavy metals to the environment.

This study was aimed at establishing a baseline concentration of lead and cadmium in soil samples from the Kamukunj and Bondeni *Jua Kali* sites and from the environs of the Naivasha garage, and has shown that *Jua Kali* enterprises are the main point sources of lead and cadmium pollution in the sample matrix. The levels of cadmium and lead in *Jua Kali* soils were quite high, with concentrations up to 70 times those found in soil samples from unpolluted areas. Such metal ions leach and find their way into water resources used by a large population for domestic and irrigation purposes. This calls for urgent regulatory policies, which will regulate the deposition of wastes away from water resources and residential areas.

The most significant statement concerning the results reported in this paper is that *Jua Kali* metallurgical enterprises are a major source of pollution of the environment by heavy metals. These findings add to the rapidly growing list of sources of pollutants which must be explored further to reduce its possible menace in the future. The data highlight to the government and the public that the levels of heavy metals released into the environment by these enterprises are too high, and that regulatory measures should be considered.

A continuous monitoring scheme of all effluents/emissions should be implemented to account for the inputs of heavy metals into the environment. A policy regulating the location of such sites should be formulated as well as the regulatory measures for disposing the wastes emanating from such sites.

References

1 Republic of Kenya (1988). Small Enterprise and *Jua Kali* Development in Kenya. Sessional Paper No. 2 of 1992, Gov. Printer. Nairobi: pp. 1-10.

2 D. Friedman (2008): Drinking Water Contaminant Levels – New York State Maximum Allowed: Building and Environmental Inspection, Testing, Diagnosis, Repair, and Problem Prevention Advice: http://www.inspect-y.com/water/levels.htm.

3 EBI (2005): Environmental Bureau of Investigation: Cadmium Levels in Drinking Water. http://www.eprf.ca/ebi/contaminants/cadmium.html.

4 International Labour Organization (ILO) (1972): Employment, Incomes and Equality – a Strategy for Increasing Productive Employment in Kenya. ILO, Geneva.

5 Environmental Data Report (1993-94): United Nations Environmental Program, GEMS Monitoring and Assessment Research Center, London, U.K., pp. 5-8.

6 A. Kabata-Pendias and K.P. Henry (1991): 'Ecological consequences of arsenic, cadmium, mercury and lead enrichment in European Soils'. In Lead, Mercury, Cadmium and Arsenic in the Environment, SCOPE 32, T.C. Hutchinson, K. Meema and C. Gordon, eds., J. Wiley, pp. 36-77.

7 A. Kabata-Pendias and K.P. Henry (1986): 'Bioavailability of Heavy Metals in the Vicinity of a Copper Smelter'. Trace Elements in Soil and Plants, pp. 211-217.

8 Republic of Kenya (1989): "Focus on *Jua Kali* Informal Sector", Gov. Printer: Nairobi.

9 R. Impens, J. Fagot, C. Avril (1991): 'Gestion des Sols Contaminés par des Métaux Lourds'. Association Française Interprofessionelle du Cadmium, Paris, France.

10 F.D.O. Oduor (1992): 'A study on the Extent to which Lead and Cadmium have Contaminated Roadside Soils and Vegetation in Nairobi and Kiambu, Kenya'. Master's Thesis, University of Nairobi.

11 S.O. Wandiga, I.O. Jumba and F.D. Oduor (1996): 'Evaluation of the Mineral Status of Grazing Cattle in Trans Nzoia District', Preliminary assessment of data, Proc. Kenya Chemical Society, Second International Conference, 6th Sept., Moi University.

12 G.N. Kamau, F.D.O. Oduor and A.N. El-Busaidy (1994): 'Use of Roadside Soils and Vegetation in Monitoring Lead Pollution from Vehicle Exhausts: A Case Study in Nairobi and Kiambu, Kenya'. Proc. Kenya Chemical Society, Inaugural Conference, 1-11 June, 1993, University of Nairobi.

13 S.J. Kimsop, G.N. Kamau and S.P. Karuu (2002): 'Analysis of Heavy Metal Pollutants along Nairobi River'. A paper presented at Kenya Chemical Society 4th Conference, Egerton University, Kenya.

AN ASSESSMENT OF EFFLUENT TREATMENT TO REDUCE HEAVY METAL LEVELS IN EFFLUENT FROM A BATTERY MANUFACTURING CONCERN

E. Muleya and F. Chiumira

Midlands State University, Department of Chemical Technology, Senka Road, P. Bag 9055, Gweru, Zimbabwe

1 INTRODUCTION

The study area is one of the point water pollution sources for chemical waste because this is a battery manufacturing concern. The city of Gweru is currently not monitoring the heavy metal levels in industrial waste, except for biological oxygen demand and pH. Heavy metals are bio-accumulative and hence are a health threat to humans, and to micro-organisms and aquatic life in general. The battery manufacturing concern uses both borehole water and tap water in its processes. In this study, borehole water was also analysed in order to assess whether pollution at the surface was percolating downwards. Tap water was used as a control because the dam from which municipal water is drawn lies in an area far removed from pollution sources. Acceptable limits for various metals in effluent discharge are as follows:

Metal	Zimbabwean Environmental Management Regulations, 2007 (ppm)	US Environmental Protection Agency, 2003 (ppm)
Fe	0.3	0.3
Pb	0.05	0.05
Cd	0.01	0.005
Mn	0.1	0.05
Zn	1.0	5

2 MATERIALS AND METHODS

2.1 Area of study

The area of study was a battery manufacturing concern in the heavy industrial sites of Gweru, Zimbabwe, along Bristol Road.

2.2 Sampling

Sampling was carried out twice a month (mid month and month end) for the three-month period July to September 2006. Five samples were collected during each sampling trip from the points that convey effluent from the factory. Tap and borehole water samples were also collected at the same time. Samples were collected in clean bottles and stored in a refrigerator before analysis.

2.3 Experimental

Apparatus: Stoppered glass bottles, Whatman 91 filter papers, appropriate laboratory glassware. Reagents: All reagents used were of analytical reagent (AR) standard. The following reagents were used: sodium hydroxide, sulphuric acid, zinc metal, iron wire, lead metal, cadmium foil, hydrochloric acid, nitric acid, distilled water, deionised water. Instrumentation: a pH meter Metrohm 744 and a flame atomic absorption spectrophotometer (FAAS) Avanta GBC 908BT version 1.32 were used after appropriate calibrations. Experimental procedure: Effluent samples were filtered to remove insoluble contaminants and a volume of 50 ml was set aside for raw effluent analysis. A volume of 200ml was put into a beaker and was adjusted to pH 6 by the dropwise addition of 0.1 M sulphuric acid in order to precipitate out zinc and copper ions. The precipitate was left to settle for about 10 minutes and then filtered off using Whatman 91 filter paper. The filtrate was further treated by the slow addition of 0.5 M sodium hydroxide from a burette with stirring to adjust the solution to pH 7, 8 and 9 to precipitate hydroxides of Fe^{2+}, Pb^{2+}, Cd^{2+}, and Mn^{2+}. The procedure was repeated using solid lime in place of sodium hydroxide solution. The treated effluent was also set aside for FAAS analysis.

2.4 Preparation of Standards

Pure metals were dissolved in concentrated acid solution to produce 1000ppm stock solutions for standard preparation. The standard solutions used were as follows:

Element	Fe	Cd	Pb	Zn	Mn
Standard/ppm	10-50	0.5-2.5	4-20	500-2000	6-30

Standards were analysed first in order to produce a calibration curve. This was then followed by analysis of treated and untreated effluent samples.

3 RESULTS

Table 1 shows the variation in the levels of heavy metals in effluent samples collected in the morning and afternoon. EWA and EWB represent morning and afternoon samples, respectively. Table 2 shows the effect of treatment of the effluent with sodium hydroxide during the period of study, with EWA and TEA representing the levels in ppm before and after treatment, respectively. ND represents non-detectable levels by the spectrophotometer. Table 3 shows results obtained when raw effluent, tap water, bore hole water and treated effluent were analyzed for the levels of heavy metals under study. The results show that there are very high levels of zinc in the borehole water. Table 4 shows the analytical results obtained after raw effluent was treated with sodium hydroxide and then compared to the analytical results obtained when lime was used instead of sodium hydroxide in treating the raw effluent.

Table 1 *Variation of heavy metal levels (ppm) in effluent in the morning and afternoon*

Date	14/07/06	28/07/06	16/08/06	30/08/06	14/09/06	28/09/06
EWAFe	1.99	0.7	ND	0.75	1.06	0.4
EWBFe	2.55	0.25	0.67	0.95	0.54	0.36
EWAPb	ND	ND	ND	ND	ND	ND
EWBPb	ND	ND	ND	ND	ND	ND
EWACd	0.38	0.02	0.03	0.07	0.14	0.06
EWBCd	0.4	0.02	0.06	0.08	0.07	0.06
EWAZn	426	256	60	133	196	148
EWBZn	444	227	119	195	215	186
EWAMn	18.46	7.84	1.34	7.09	2.36	2.12
EWBMn	18.39	6.823	2.35	18.89	1.51	7.7

Table 2 *Heavy metal levels (ppm) in effluents before and after treatment*

Date	14/07/06	28/07/06	16/08/06	30/08/06	14/09/06	28/09/06
EWAFe	1.72	0.77	ND	0.54	1.21	0.65
TEAFe	0.074	ND	ND	ND	ND	ND
EWAPb	ND	ND	ND	ND	ND	ND
TEAPb	ND	ND	ND	ND	ND	ND
EWACd	0.38	0.02	0.03	0.08	0.15	0.07
TEACd	0.06	ND	ND	ND	ND	ND
EWAZn	407.3	249.5	70.18	144.29	177.26	143.87
TEAZn	98.33	58.57	ND	ND	ND	ND
EWAMn	16.04	7.79	1.33	6.45	2.37	1.96
TEAMn	0.22	ND	ND	ND	ND	ND

Table 3 *A comparison between heavy metal levels in tap water, borehole water and treated effluent during the period of study*

Sample	Fe	Pb	Cd	Zn	Mn
TPA1	0.54	ND	0.03	ND	ND
BHA1	0.57	ND	0.04	12.04	ND
TEA1	0.08	ND	0.06	106.92	0.07
TPA2	0.22	ND	0.02	3.00	ND
BHA2	0.18	ND	0.02	10.08	ND
TEA2	ND	ND	0.01	40.53	0.09
TPA3	ND	ND	ND	ND	ND
BHA3	ND	ND	0.01	14.44	0.05
TEA3	ND	ND	0.02	ND	ND
TPA4	0.25	ND	ND	ND	ND
BHA4	0.42	ND	0.03	17.43	ND
TEA4	ND	ND	ND	ND	ND

TPA5	ND	ND	0.02	1.43	ND
BHA5	ND	ND	0.02	14.16	ND
TEA5	ND	ND	ND	ND	ND
TPA6	ND	ND	0.01	ND	ND
BHA6	ND	ND	0.01	13.13	0.06
TEA6	ND	ND	ND	ND	ND

Key

TPA tap water samples; BHA borehole water samples; TEA treated water samples

Table 4 *A comparison of levels of metals (ppm) in samples treated with lime to those treated with sodium hydroxide*

Sample	Fe	Pb	Cd	Zn	Mn
TEA1E	ND	ND	ND	ND	ND
TEA2E	ND	ND	ND	ND	ND
TEA3E	ND	ND	ND	ND	ND
TEAL1	ND	ND	ND	ND	ND
TEAL2	ND	ND	ND	ND	ND
TEAL3	ND	ND	ND	ND	ND
TEA1F	ND	ND	ND	ND	ND
TEA2F	ND	ND	ND	ND	ND
TEA3F	ND	ND	ND	ND	ND
TEAL2B	ND	ND	ND	ND	ND
TEAL2B	ND	ND	ND	ND	ND
TEAL3B	ND	ND	ND	ND	ND

Key

TEA(1-3)E - Treated with NaOH; TEAL1-3 - Treated with Lime

4 DISCUSSION

The results shown in Table 1 indicate that zinc is the major heavy metal pollutant in the effluent, ranging in concentration from about 50-450ppm. This is because zinc chloride is used as an electrolyte, and as a raw material to produce cans and plates in the plant. Zinc levels in the effluent were far above the recommended limit of 1ppm according to the Zimbabwean wastewater standards. The zinc levels were generally higher in the afternoon as compared to the morning samples, which could be explained by the fact that production in the plant would be at its peak in the afternoon. The results shown in Table 1 also indicate that manganese levels are high, with concentrations ranging from 1-18ppm. This is far above the recommended limit of 0.1ppm. Manganese is used as manganese dioxide in the battery manufacturing industry to make an electrolyte paste with other components. Cadmium levels in the effluent, i.e. 0.02-0.4ppm (Table 1), are also high, and are above the recommended level of 0.01 ppm. Cadmium is used in very low quantities in the zinc alloy smelting in the plant. Lead was not detected during the period of study. Lead is however also used during the formation of the zinc alloy. This showed that the plant operation uses lead efficiently without it being lost into the environment. Iron is not a raw material in battery manufacturing, but was detected during this study at concentrations up to 2.50ppm. Recommended limits for iron are 0.3ppm. The presence of iron pollution could be due to rusting of equipment and machinery in the plant and chemical corrosion by detergents during clean-up of utensils.

Treatment of the raw effluent with both lime and sodium hydroxide reduced the levels of all the metals, and even zinc was reduced to non-detectable levels in some cases (Table 4). A comparison of heavy metal levels in tap water, borehole water and treated water (Table 3) showed that borehole water had high concentration of zinc ions (10-17 ppm). Tap water mostly showed non-detectable levels of the ion. This shows that the surface pollution ends up with zinc being leached into underground water. This type of pollution is disturbing and deserves attention. Treated effluent had reduced zinc levels from non-detectable levels to 98ppm (Table 2). The higher levels of zinc could be explained by incomplete precipitation that could have taken place. Further and repeated precipitation procedures can reduce the zinc levels further.

5 CONCLUSION

The battery manufacturing company is producing effluent with above permissible limits of the heavy metals for industrial effluents. The lime and sodium hydroxide precipitation methods reduced heavy metal levels in the effluent. Zinc and manganese were the major pollutants to the environment. Manganese pollution is not only a danger to human health. For example, water treatment works may have problems associated with the presence of "manganese" bacteria, which concentrate manganese, and may give rise to taste, odour and turbidity problems in the distributed water.[9] Oxidation of manganese ions in solution results in the precipitation of manganese oxides and incrustation problems. Even at concentrations of approximately 0.02 mg/L, manganese may form coatings on water distribution pipes that may slough off as black precipitates.[9] At concentrations exceeding 0.15 mg/L, manganese stains plumbing fixtures and laundry and causes undesirable tastes in beverages.[9] Cadmium levels were also high enough to warrant the attention of the waste water regulatory organizations because it is a cumulative poison and causes irreversible damage to the human body. Lead was not detected throughout the study. This shows that there is efficient use of lead in the plant. Borehole water had high levels of zinc, showing that zinc leaching is taking place and polluting Gweru's underground water.

Acknowledgements

I wish to thank Midlands State University, Zimbabwe, for providing the funds to undertake this research. I would also wish to thank the battery manufacturing concern for allowing us to take samples for this research in their premises, and for allowing us the use of their laboratory.

References

1 G. Hill, *Chemistry Counts*, Hodder and Stoughton, London, 1986.
2 EPA 816-F-03-016(2003), http://www.epa.gov/safewater/contaminants/index.html.
3 B.J. Alloway and D.C. Ayres, *Chemical Principles of Environmental Pollution*, Blackie Academic, London, 1993..
4 *Hach Water Analysis handbook*, Hach Company, Colorado, 2nd Ed., 1992, p. 35.

5 L.M. Haddad, M.W. Shannon and J.F. Winchester, *Clinical Management of Drug Overdose*, Saunders, Philadelphia, 3rd Edition 1998.
6 National Institutes of Health (NIOSH) USA, http://www.ods.od.nig.gov.
7 The Analyst (2008), http://www.digitalnaturopath.com/cond/C512265.html
8 Springboard (2008), Nutrition on Line: http://www.springboard4health.com/notebook/min_manganese.html.
9 Canadian Council of Resource and Environment Ministers (CCREM), Canadian water quality guidelines, prepared by the Task Force on Water Quality Guidelines, 1987.

SPECTROSCOPIC DETECTION OF GASEOUS AMMONIA IN THE ENVIRONMENT

C.B.O. Kowenje[1] and D.C. Doetschman[2]

[1] Department of Chemistry, Maseno University, Private Bag Maseno, Kenya.
ckowenje@maseno.ac.ke (corresponding author).
[2] Department of Chemistry, SUNY Binghamton University, NY, USA.
ddoestch@binghamton.edu

1 INTRODUCTION

Ammonia pollution is mainly from agricultural sources (especially animal manure, forestry and fertilizers) and it causes damage to sensitive habitats, affecting plants and biodiversity including soils and water, and upon accumulation in the atmosphere leads to long-range acidification.[1,2] Pollution by ammonia is also a recognized problem in mines and in the marine environment.[2] The emissions of ammonia are somewhat more uncertain than those of SO_2 and NOx. However, ammonia synergistically interacts with the two gases in the atmosphere leading to little overall impact of their legislative control of the SO_2 and NO_x in several countries.[2] To date, little action has been taken by environmentalists to control the emission of ammonia, despite the fact that, ironically, upon decomposition in the environment, it has a similar fate to that of NO_2.[3]

Therefore, without improvising how to effectively measure the ammonia emissions, there may be no impetus for measures to control it. The current methods of monitoring environmental ammonia, such as polymer films and use of the Kjeldahl method, are less sensitive and very expensive and therefore preclude the main producers of ammonia, the poor rural farmers, from monitoring their share of ammonia emission.

Spectroscopy is a very sensitive method and has been successfully used in wet chemistry to identify, characterize, and quantify anion-cation ligation interactions.[4] Differences in geometry, coordination, and the type of ligand or ligands around a central atom or ion is reflected in the intensity and energy of the spectral transition of the central ion. This is because electronic energy changes as presented by 10 Dq (Δ_o) in octahedral symmetry are determined by the coordination chemistry of the central metal and by the nature of ligands around it.[4] The spectrochemical series of the ligands gives rise to different 10 Dq splittings of transition metal cations. Normally, the changes in electric field splitting like the Δ_o changes in solid samples are monitored by diffuse reflectance spectroscopy (DRS) and the spectra are recorded in Kubelka-Munk units (kmu).

The ability of ligands to give rise to different crystal field splittings is expressed relatively in the ligand spectrochemical series, which for the ligands considered in this work is water $\leq$ acetone < pyridine < ammonia.[5,6] Thus ammonia is the strongest ligand of them all and is expected to strongly bind to the cations exchanged in the zeolites.[6] The series gives rise to different wavelengths for a d-d transition in different ligand–metal cation combinations.

The infrared (IR) vibration frequencies of complexes formed in cation-exchanged zeolites are categorized into two main regions. The 4000 – 1000 cm^{-1} region is associated with ligand vibrations in which any changes noticed in the IR vibration band of the region, as compared to the bands of the free molecular ligand, may provide information on the perturbations caused by the cation-ligand interactions.[7,8] The region below 1000 cm^{-1} provides information about the structure of the M^{n+} coordination sphere, the nature of the metal-ligand (M-L) bond and modes the zeolite structure.[8] Thus M-L vibrations would be affected by the strengths, covalency and ionicity of the bonds formed and the reduced mass of the entities involved.[9] The above effects on the vibration frequency, insofar as they affect the vibrational force constant k and the effective reduced mass of the vibration, are quantitatively represented by the Equation (1) below for the case of a diatomic system.

$$\bar{\nu} = \frac{1}{2\pi c}\sqrt{\frac{k}{\mu}} \quad cm^{-1} \tag{1}$$

where $\bar{\nu}$, c, μ, and k are the fundamental vibration frequency, velocity of light in a vacuum, reduced mass, and the bond force constant respectively.

Therefore, such ligand-specific transitions may be used to identify a particular ligand or distinguish one from another, when other conditions are fixed. As a result, the presence of certain ligating agents in the environment may be detected by looking for the changes associated with their Cu(II) or Co(II) complex spectra when Cu(II) or Co(II) exchanged zeolites are exposed to the ligands.[6,7] Upon exchange into the zeolites (see Figure 1), the cations occupy the sites numbered I to III. Within these sites ligations with adsorbates are possible.

The resulting coordination of the exchanged cation and the zeolites framework also affects the framework IR vibration band positions.[4] Thus the groups of single four rings (S4R), single six rings (S6R), and double six rings (D6R) shall have distinct vibrational bands.[8] The relative ligand strengths are subtly reflected in their crystal field splittings and in the energies of their relative IR vibration modes.[8]

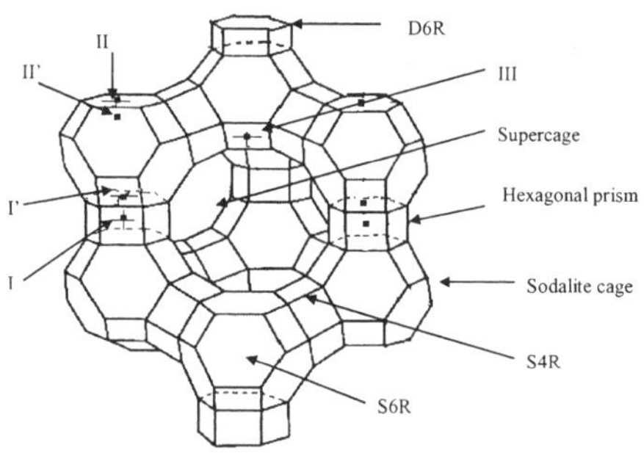

Figure 1 *Faujasite-X cages and the cation exchange sites (I – III). Where S4R, S6R, and D6R stand for single-4 -ring, single-6-ring, and double-6-ring structures respectively.*

Cu(II) and Co(II) were the first cations to be chosen for the study because they form complexes that are easy to study by DRS and IR methods. The work may be employed in qualitative monitoring of these adsorbates or their related functional groups in the environment. The characteristic electronic and bonding behavior for particular ligand-metal ion systems could be used by incorporation of tablets of variously exchanged zeolites into a multi-ligand array detector or sensor for the ligand agents.

2 EXPERIMENTAL SECTION

1.1 Materials used

Faujasite-X zeolite (NaX) (Si/Al = 1.23, *ca.* 2 μm particle size, from Aldrich Chemical Company), ammonium hydroxide (assay 29+%, from Fisher Chemical Company), cobalt(II) chloride (99+ % assay) and copper nitrate hydrate [$Cu(NO_3).^5/_2H_2O$, 101.7 % by EDTA complexation, from J.T. Baker Chemical Company] were used.

1.1.1 Preparation of the Cation Exchanged Faujasite Samples. The required concentration of $CoCl_2$ or $Cu(NO_3).^5/_2H_2O$ in the ratio of 1 g of zeolite to 20 ml of solution were mixed in an Erlenmeyer flask. The mixture was refluxed for 72 hours (h.) at *ca.* 90°C. It was then allowed to cool to room temperature after which time the supernatant solution was decanted. The residue was centrifuged at 3500 rpm for 30 minutes. Then the residue was washed free of nitrate ions four times with deionised water at 10 ml of water per gram of solid and centrifuged again. The exchanged zeolites were then air-dried in an oven at 60°C for 48 hours, in order to remove most of the non-ligand water. The required amount of the desired exchange concentration of the 60°C dried CuX was put in the sample tube and dehydrated by raising the temperature by 50°C every 30 min to 450°C under 10^{-5} torr vacuum conditions. The sample was then held at that temperature for 4 h. after which time it was allowed to cool and then used as such. The resulting samples were characterized by atomic absorption spectrometry with a Perkin Elmer AAS-Analyst 300 spectrometer in order to determine the amount of copper exchanged into the zeolite.

1.1.2 Ligation of the Cation Exchanged Zeolites. After the required amount of ammonia (also referred to as an adsorbate) was put in the liquid transfer storage tube, which was in turn connected via an inter-connecting valve to a sample tube containing cation exchanged zeolite samples, the ammonia was frozen and thawed repeatedly until no further pressure changes were detectable. The inter-connecting valve to the sample tube was then opened and the adsorbate vapor allowed to vaporize over to the sample. The set-up was then allowed to stand for *ca.* 8 h. at room temperature for complete and more homogeneous adsorption of the adsorbate into the exchanged zeolite. The ammonia-copper exchanged zeolite mixture was then warmed in a 60°C water bath to achieve further homogeneity in the distribution of the ammonia in the Cu/Co-zeolite host. The above procedure was repeated for acetone, pyridine and water for comparison with the ammonia data.

1.2 **Experimental methods**

1.2.1 Diffuse Reflectance Measurements. Adequate amounts of the sample were packed in sealed quartz glass sample cuvettes and the diffuse reflectance spectroscopy (DRS) measurements were made from 200 to 1100 nm at a resolution of 2 nm using a Perkin-Elmer Lambda 2S UV-Vis spectrometer at room temperature. A $BaSO_4$ disc was used as the reference standard. The dehydrated zeolite samples were initially transferred under an argon atmosphere in a glove box in order to exclude ambient humidity from the samples and the cuvettes were filled and sealed in the glove box.

1.2.2 Infrared Spectral Measurements. A mixture of *ca.* 1% sample and *ca.* 99% KBr was ground in a glass mortar to fineness. The weighing and grinding were both performed under an inert (argon gas) atmosphere in the glove box. Then *ca.* 0.10 gram of the ground mixture was pressed at 10 Ton pressure for *ca.* 10 minutes. The resulting pellets of average wt 0.0946 ± 0.0002 grams and width 0.0325 ± 0.0013 cm were then fixed in a FT-IR Bruker Equinox 55 spectrometer at a nominal resolution of 2 cm^{-1}. However, the actual amount of the exchanged zeolite materials was adjusted for each pellet, depending on how much adsorbate was exposed to the sample. A total of 128 scans were collected for each sample spectrum. The spectrometer was purged with nitrogen gas for 30 minutes before and after pellet insertion, after which time the spectrum was recorded over the 4000–400 cm^{-1} range.

3 RESULTS AND DISCUSSIONS

1.1 Diffuse Reflectance Measurements

In Figure 2 below, the λ_{max} for all ranges of copper per Unit Cell (Cu/UC) samples peaked at 908 nm. Electronic transitions, an example of which is indicated by the DRS peaks in this case, are metal-ligand specific,[4] meaning that up to 24 Cu/UC, all the exchanged Cu^{2+} ions reside in similar chemical environments in the zeolite matrix. The consistency in the λ_{max} values implies that any Cu/UC exchange levels, except that at 38 Cu/UC, can be used for the study.

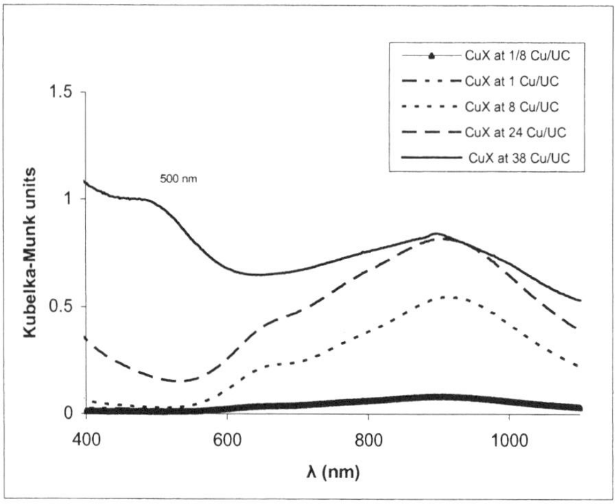

Figure 2 *DRS of CuX at different concentration levels of Cu/UC.*

The λ_{max} peaking at approximately 500 nm indicated a different chemical environment and therefore the 38 Cu/UC may not be suitable for this study as the peak could introduce some interference. Since the highest sensitivity was shown by the 24 Cu/UC, it shall be used in all subsequent studies.

Figure 3 shows a significant wavelength shift and intensification of the peaks in the visible region upon addition of ammonia to the sample. Addition of ammonia to the dehydrated CuX (Figure 3) reduced the intensity of the λ_{max} at *ca.* 908 nm and the λ_{max} position was shifted to *ca.* 630 nm. The shift was accompanied by a change from a greenish-yellow to a bluish color. The intensification and the shifting of the λ_{max} are consistent with a less symmetric complex being formed.[4] In this case, it shows ammonia is reaching the copper ions. Such transitions are ligand-metal specific[12] and can therefore be used as a fingerprint for the ammonia-copper complex.

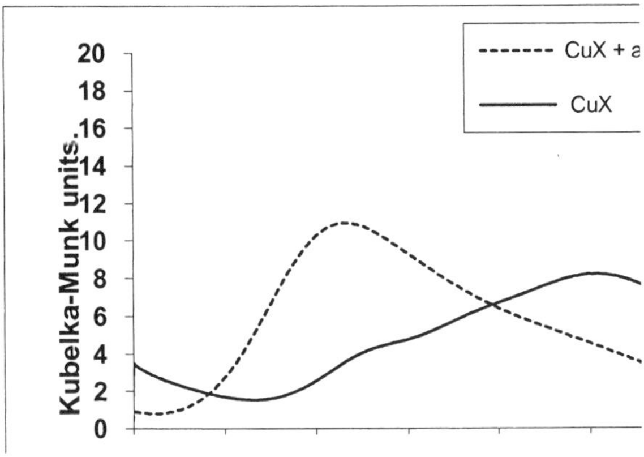

Figure 3 *Diffuse reflectance spectra of 24 Cu/UC CuX before and after ammonia addition*

The amount of Cu/UC in the zeolites matrix did not interfere with the peaking position after the ammonia exchange (see Figure 4). Figure 4 shows a consistent increment in the intensity of the peak at 630 nm with the increase in both the amount of Cu/UC and the ammonia added. According to Delabie *et al.*,[11] such a consistency is a manifestation of a specific metal-ligand interaction.

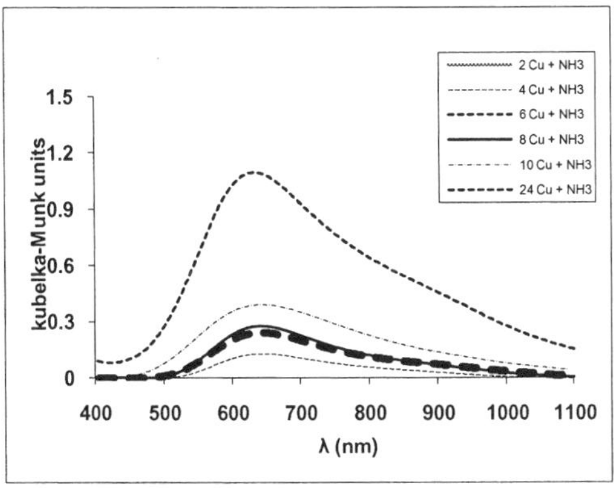

Figure 4 *Diffuse reflectance spectra of CuX at various Cu/UC after ammonia addition.*

To confirm that the specific DRS transitions are not unique to CuX, the same considerations were performed for Co/UC samples and the results shown in Figure 5. The λ_{max} for all the ranges of the Co/UC were at ca. 590 nm. The increase in the intensity of the λ_{max} at ca. 590 nm with increasing amounts of Co^{2+} ions exchanged in is consistent with similar coordination for the resulting complex at a lesser symmetric site.[7] Similarly, as was the case with Cu/UC, all the ranges for CoX up to 24 Co/UC may be used for the study. When the CoX at 24 Co/UC was exposed to ammonia, the result in Figure 6 was obtained.

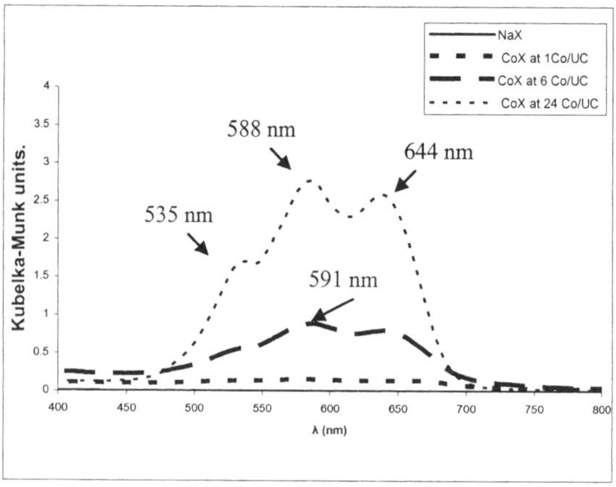

Figure 5 *Diffuse reflectance spectra of dehydrated CoX at various Co/UC at 24 Co/UC.*

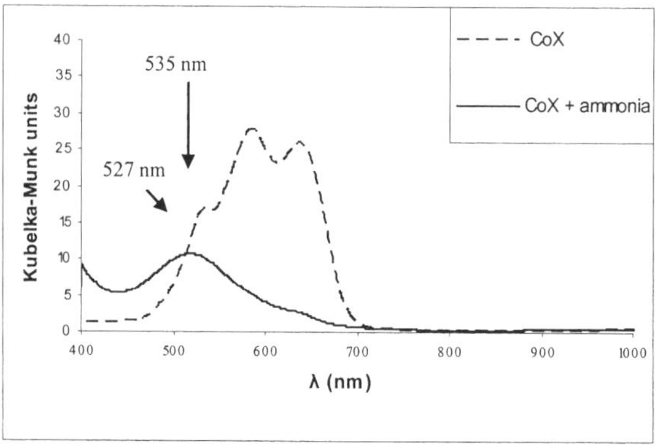

Figure 6 *IR of CoX at 23 Co/UC before and after ammonia addition.*

The shift in λ_{max} (Figure 6), after exposure to ammonia vapour, is an indication that the ammonia has interacted with the Co^{2+} exchanged into the zeolite.[4] To exclude possible interferences from other environmental contaminants, the study included additional adsorbates with similar functional groups (such pyridine), abundant in the ambient atmosphere (such as water), and abundant in organic environments (such as acetone), and the results expressed in cm^{-1} energy units shown in Table 1.

Spectroscopically, different ligand adsorbates show fingerprint DRS peaks and the peaks can be used to identify the individual molecules.[13,14] However, differences in the energy values from the table is noteworthy.

Table 1 *The changes in λ_{max} shifts from dehydrated 24 Cu/UC for the various adsorbates in cm^{-1} energy units (The assumption is made that the resulting complexes, after exposure to the adsorbates, have comparable geometry).*

Sample	Changes in λ_{max} shifts from 908 nm (converted to cm^{-1})
Ammonia	4860
Pyridine	3957
Acetone	1807
Water	1661

The λ_{max} for ammonia is statistically different from the rest. The difference shows that ammonia is a stronger ligand compared to the rest of the adsorbates[13,14] and can therefore be detected, without interferences, in the presence of the other adsorbates.

1.2 Infrared measurements

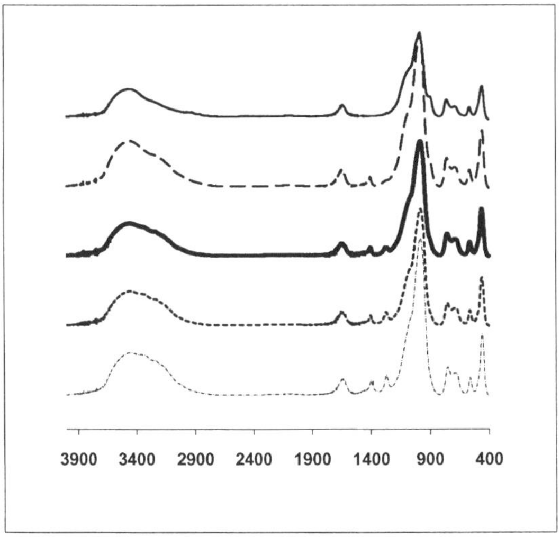

Figure 7 *Infrared spectra of the 24 Cu/UC CuX exposed to different amounts of NH_3 per Cu^{2+} exchanged in CuX. {(a) 0; (b) ½; (c) 1; (d) 2; and (e) 4 NH_3 per Cu/UC}.*

The measurements will concentrate in the 2000–1000 cm^{-1} region for ligand vibrations and the region below 1000 cm^{-1} for the metal-ligand (M-L) bands.[15] Upon Cu^{2+} exchange, the Cu-Zeolite (Cu-Zeo) symmetric vibration band forms at 896 cm^{-1} (see Figure 7a).[16] In Figure 7, the Cu-Zeo band at 896 cm^{-1} is seen to disappear when the CuX is subsequently exposed to ammonia gas.

In addition, Figure 7 shows the concomitant relative rise of the band at *ca.* 1270 cm^{-1} and the demise of the shoulder at 896 cm^{-1} with the increase in the amount of ammonia gas applied to the dehydrated 24 Cu/UC. The proportionate increase in the relative intensities of the 1270 cm^{-1} band with the amount of ammonia absorbed into the dehydrated CuX indicates that ammonia directly controls this mode. Literature values for Hexamine-Copper Complex occurs in this neighbourhood.[17] Thus we can assign the mode to Zeo-Cu(-NH_3) symmetric vibrations. As such, the accompanying decrease in the relative intensity of the 896 cm^{-1} band is attributable to a reduced interaction of the Cu-Zeo bonding as ammonia removes the Cu^{2+} from the zeolite framework,[18] according to Equation 2 below. Therefore, IR measurements may offer semi-quantitative evaluation of the amount of ammonia in the environment.

$$Cu(O_{zeo})_x + yNH_3 \rightarrow Cu(O_{zeo})_{x-y}(NH_3)_y + \ldots \qquad (2)$$

When CoX was similarly treated with ammonia vapour a band at 1312 cm^{-1}, similarly associated to Co-NH_3 symmetric vibration, was observed (see Figure 8). When compared to other adsorbates (pyridine, acetone and water vapour) for their ability to shift the Metal-Zeolite bands (at 896 cm^{-1} for CuX and 918 cm^{-1} for CoX), the band positions for each of the metal-adsorbate vibrations was distinct (see Table 2).

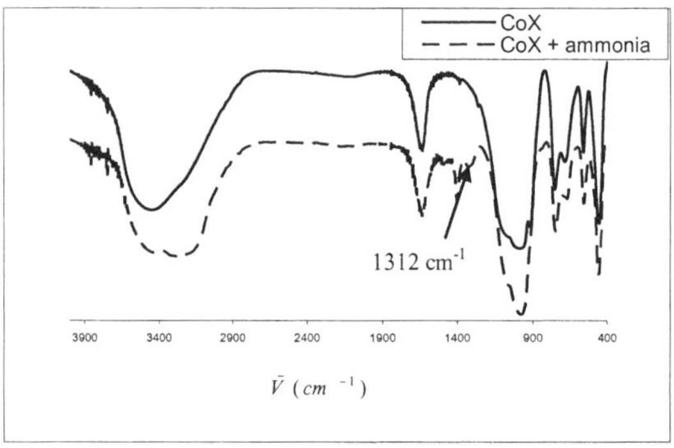

Figure 8 *Infrared spectra of dehydrated NaX and 24 Co/UC CoX samples.*

A germane attempt to rank adsorbate ligands in terms of their Cu^{2+} d-d splitting when the cation (Cu^{2+}) is coordinated on a silica surface had earlier been made by Trouillet *et al.*, in a Cu/SiO_2 matrix.[19] Trouillet and co-workers applied the rule of 'averaged environment' to conclude that the order was ethylenediamine (en) > ammonia > water > silica (SiO_2). A series similar to this is obtained from the present study, but includes pyridine and acetone. Thus, from Table 2 and Figure 7, we can therefore conclude that it is possible to use IR spectroscopy to monitor ammonia in the environment.

Table 2 *Infrared fingerprint bands associated with the various adsorbates; ammonia, pyridine, acetone, and water when separately exposed to 24 Co/UC and 24 Cu/UC (designations w and s refer to weak and strong bands respectively).*

Adsorbate	**24 Co/UC**	**24 Cu/UC**
Ammonia	1312 s	1270 s
Pyridine	1360 w	1250 s
Acetone	1400 s	1350 w
Water	1447 s	1280 s

4 CONCLUSIONS

This work shows the spectroscopic characteristics of ammonia in the presence of Cu^{2+} and Co^{2+} cations exchanged in the Faujasite-X zeolite and its ligand strength. This data may be used to identify and semi-quantify the presence of ammonia in the presence of pyridine, acetone and water. Specifically, both DRS and IR need to be used together in any attempt to formalize the sensoring of ammonia in the environment. Further research is, however, needed to quantify ammonia in the environment.

Acknowledgements

The authors kindly acknowledge the help, both financial and in terms of materials received, from the Department of Chemistry, SUNY Binghamton University, New York. In addition, the staffs both in Prof. D.C. Doetschman's research group and in the Department of Chemistry, Maseno University, Kenya, are equally saluted.

References

1 A. Spangenberg, H. Utischig, T. Preuhsler and H. Pretzsch, *Plant and Soil*, 2004, **262,** 1-2, 337.
2 M. Perez-Soba, L. Van Der Eerden and I. Stulen, *Plant Physiology and Biochemistry*, 1994, **32,** 539.
3 Environmental monitoring measurements available at www.pmeasuring.com.
4 B.N. Figgis and M.A. Hitchman, *Ligand field theory and its application,* Wiley-VCH, New York, 2000, pp 1-26.
5 H.L. Kwang, and P.G. Clare, *J. Am. Chem. Soc.,* 2000, **122**, 9768.
6 I.E. Maxwell and J.J. De Boer, *J. Phys. Chem.,* 1975, **79**, 1874.
7 G.L. Miessler and D.A. Tarr, *Inorganic Chemistry 2nd Ed.,* Prentice-Hall, New Jersey, 1998, pp 352-383.
8 V.Y. Borovkov, M. Jiang and Y. Fu, *J. Phys. Chem. B,* 1999, **103**, 5010.
9 T. Ichikawa and L. Kevan, *J. Phys. Chem.*, 1983, **87,** 4433.
10 C.O. Kowenje, Spectroscopic characterization of the metal cation siting and the adsorbate-cation interactions in copper(II) and cobalt(II) exchanged Faujasite-X Zeolite, PhD thesis; SUNY Binghamton USA, 2006, available at www.il.prorequest.com (UMI number: 3220349).
11 A. Delabie, K. Pierloot, M. Groothaert, B.M. Weckhuysen and R.A. Schoonhydt, *Microporous and Mesoporous Materials,* 2000, **37**, 209.
12 H.S. Lee and K. Seff, *J. Phys. Chem.*, 1981, **85**, 397.
13 J. Barrett *Understanding Inorganic Chemistry, The underlying physical principles.* Ilis Horwood, New York, 1991, pp 186-224.
14 P.J. Durrant and B. Durrant, *Introduction to Advanced Inorganic Chemistry,* John Wiley and Sons, NY, 1970, pp 967-986.
15 C. Lamberti, S. Bordiga, M. Salvalaggio, G. Spoto, A. Zechina, F. Geobaldo, G. Vlaic and M. Bellatreaia, *J. Phys. Chem.,* 1997, **101**, 344.
16 J.R. Dwyer, V. Parish and N.P. Evmerides, *Inorganic Chemical Acta*, 1983, **75**, 229.
17 N.B. Colthup, L.H. Daly and S.E. Wiberley, *Introduction to infrared and raman spectroscopy,* 3rd Ed., 1990, Academic Press, Inc., NY, pp 21-89.
18 B. Su, V. Norgerg and J.A. Martens, *Langmuir*, 2001, **17**, 1267.
19 L. Trouillet, T. Toupance, F. Villain and C. Louis, *Phys. Chem. Chem. Phys.*, 2000, **2,** 2005.

JATROPHA CURCAS L. AS A SOURCE FOR THE PRODUCTION OF BIODIESEL IN KENYA

A.W. Wagutu,[1] S.C. Chhabra,[1] C.L. Thoruwa,[1] T.F. Thoruwa[2] and R.L.A. Mahunnah[3]

[1] Chemistry Department, Kenyatta University, P.O. Box 43844-00100, Nairobi, Kenya
[2] Engineering Departments, Kenyatta University, P.O. Box 43844-00100, Nairobi, Kenya
[3] Traditional Medicine Research Institute, Muhimbili University Hospital, P.O. Box 65000, Dar es Salaam, Tanzania

1 INTRODUCTION

The commercial energy sector in Kenya is dominated by petroleum and electricity as the main energy sources of the modern sector of the economy. Nevertheless, petroleum rates second (22%) to biomass in terms of total national energy consumption, while electricity accounts for only 9%. Kenya is still heavily dependent on wood fuel and other biomass fuels which together account for 68% of the total energy consumption.[1]

However, as energy and the economy are closely linked, it is clear that a growing Kenyan economy will demand much higher levels of energy consumption in the form of liquid fuel, especially in transport and communication, and the industry and electricity generation sectors. The transport and communication sector recorded a real growth of 8.3% in 2005, while industry and generation of electricity demand grew by 5.2% in 2006. The demand is expected to rise even more as vision 2030 approaches. In addition, the dire need to protect the remaining vegetation cover requires substitution of fuel wood by liquid or gas fuels. Increased environmental awareness and the constrained world's petroleum supply also shifts attention towards exploitation of alternative fuels for diesel engines.

Vegetable oils from crops, with a total lipid content of at least 20% by weight, such as soy bean, rape seed and sunflower, have been studied and cultivated in plantations as sources of feed stock for biodiesel production in the USA and in European countries such as France, Austria and Italy. For Kenya and Africa at large, where there is a dire need for food, the use of such edible oil crops for fuel would be ridiculous. *Jatropha curcas* (family Euphorbiaceae), which constitutes 53-57% non-edible oils, has therefore been evaluated as a source of oil for biodiesel production.

The preparation of *Jatropha curcas* methyl esters (JCME) and a study of their chemical and fuel properties are presented in this paper. The results are compared with those of petroleum diesels (Diesel No.2 and kerosene).

2 EXPERIMENTAL

2.1 Collection and Preparation of Plant Materials

Jatropha seeds were obtained from a test plantation owned by the Vanilla Jatropha Development Foundation (VJDF), Nairobi, and their identity was confirmed at the Kenya National Herbarium, Nairobi. Decortication to remove the hard outer shells was done manually. The seeds were then washed with water to remove dust and mud, sun-dried for three days and reduced to powder using a blender.

2.2 Preparation of the Oils

The powdered sample was soaked in *n*-hexane for three days and the oil-hexane miscella was then filtered by suction through a Büchner funnel. Most of the *n*-hexane in the mixture was removed using a rotary evaporator with the temperature of the bath set at 40 °C, and the resulting oil sample was left in a fume hood for 72 hours to allow the remaining solvent to evaporate. The oil was then treated with anhydrous Na_2SO_4 to remove all moisture. The resulting dry oil was used for biodiesel synthesis by reaction of the oil with anhydrous methanol in presence of NaOH catalyst.

2.3 Preparation of Methyl Esters (Biodiesel)

Commercial grade methanol was dehydrated by Lund and Bjerrum's method, using magnesium turnings in a reaction catalyzed by iodine.[4] Exactly 1000 ml of dry *Jatropha* oil was allowed to react with a solution of 10 g anhydrous NaOH [1% (w/v) of oil] in 225 ml dehydrated methanol (22.5% by volume of oil). The reaction was performed in a 2L QuickFit Erlenmeyer flask. A magnetic rod was placed inside the flask which was then fitted with a condenser. The flask was placed on a hot plate magnetic stirrer with the temperature set at 70 °C and the magnetic revolutions at 700 rpm. The mixture was stirred continuously and refluxed for 4 h. The solution was then cooled, placed in a separating funnel and left to stand for 12-20hrs. The mixture separated into two layers, a top ester layer and a bottom glycerol layer. The ester layer was separated and washed with warm distilled water until the wash water became clear. The ester was then dried using anhydrous Na_2SO_4 (10 g of the salt per 100 ml of the ester). The general reaction for the preparation of methyl esters is shown in Figure 1.

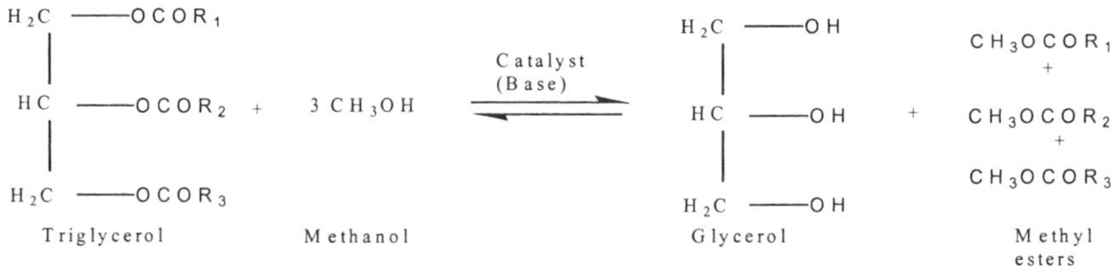

Figure 1 *Transesterification reaction.*

2.4 Oil Properties

The viscosity of the oil was determined using a *Cannon-Fenske* viscometer at 40 °C. Density and refractive index were determined using hydrometers and a refractometer, respectively, at room temperature. Acid value, iodine number and saponification value were analyzed according to the Standard Methods of American Oil Chemists.[5,6]

The iodine value of the ester was determined by Wijs' method,[10] and was also calculated from the GC-MS mixtures using Equation 1.[7]

$$IV_{Mixture} = \sum 100 \times \frac{A_f \times 253.81 \times db}{MW_f} \qquad \textbf{(Equation 1)}$$

Where IV is the iodine value, A_f is the percentage of fatty acid esters in the mixture, db is the number of double bonds, MW_f is the molecular weight of the fatty components as shown by MS spectra and 253.81 is the atomic weight of two iodine atoms that are theoretically added to one double bond.[7] The values obtained by both methods are given in Table 2.

2.5 Analysis of Fuel Parameters

The determination of fuel parameters was carried out according to standard procedures of the American Society for Testing and Materials (ASTM) for diesel fuels analysis at the Kenya Pipeline Company Limited (KPC), Nairobi, as outlined in Table 1.

Table 1 *Fuel parameters determined for the biodiesels.*

Property	Method	Apparatus
Flash point	ASTM D93	*Pensky-Martens* closed cup automated
Water content	ASTM D95	Sample distillation apparatus
Kinematic viscosity (40 °C)	ASTM D445	*Cannon-Fenske* capillary viscometer tube
Distillation	ASTM D86	Automated distillation unit
Copper strip corrosion	ASTM D130	Copper corrosion test bomb
Cetane index	ASTM D4737	-
Density	ASTM D1289	API Hydrometers
Color	ASTM D1500	Colorimeter
Calorific value	ASTM D240	Automated adiabatic bomb calorimeter
Refractive index	-	Automated refractometer

2.6 Ester Composition

Determination of the fatty acid composition of the *Jatropha* biodiesel was performed using a GC-MS method, employing HP 5307A GC equipment with a VG 12-250 mass selective spectrometer. The column was a 0.2 mm id, 50 m long and 0.33 µm thick HP ULTRA1 (methyl silicone) high performance capillary column. It was run in splitless mode under the following conditions. An initial temperature of 50 °C was held for 5 minutes to elute the solvent. The temperature was then increased at 3 ^{0}C per min up to 280 ^{0}C, with 20 minutes final time. Nitrogen was used as the carrier gas, at a flow rate of 0.84 μL/min.

Calibration was performed using a standard mixture of fatty acid methyl esters (FAME) in methylene chloride. The standard mixture contained FAME from C6-C20, from fully saturated to tri-unsaturated. A 1 μL aliquot was mixed with 100 μL of HPLC grade DCM, and a 6 μL injection was used for the analysis.

3 RESULTS AND DISCUSSION

The experimental yields of the methyl esters from *Jatropha curcas* oils, prepared as shown in Figure 1, were determined to be in the range 90.7-96.2% by volume of oil. The losses in comparison to the 100% theoretical yield were attributed to the formation of salts of free fatty acids present in the oil, which is evidenced by the high acid value of the oil (5.2) as compared 0.2 of the resulting esters (Table 2).

3.1 Properties of *Jatropha* Oil and the Methyl Esters derived from it

The properties of the *Jatropha* oil and the resultant methyl esters are given in Table 2, together with those of Diesel No.2 and kerosene fuels.

Table 2 *Properties of the methyl esters*

Sample	Iodine Value by		Acid No. (mg KOH/g)	Saponification value	Refractive Index	Calorific value (MJ/Kg)
	Wijs	GC-MS				
JCO*	96	-	5.18	112.2	1.4685	42.9
JCME*	102	106	0.24	192.6	1.4461	40.8
Diesel No.2	-	-	-	-	1.4719	42.4
Kerosene	-	-	-	-	1.4423	43.6

* JCO = *Jatropha curcas* oil; JCME = *Jatropha curcas* methyl ester

3.1.1 Iodine Value

The ASTM standard does not give restriction to the iodine value of methyl esters but the European biodiesel standard (EN 14214) states that the iodine value should be no more than 120. JCME is within the requirements of EN 14214. The iodine number of conventional (automotive) diesel fuel is reported to be 10.[2] Based on the results shown in Table 2, the iodine value of the methyl esters differs only slightly from that of their parent oil. The iodine value derived from the GC-MS chromatograms are higher than those derived from the Wijs solution method, mainly because Equation 1 assumes full iodonization and treats all the double bonds as being equally reactive to oxidation.[7]

3.1.2 Saponification Value (SV)

The saponification of the oil or of an ester has a direct relationship to the average molecular weight of the fatty compounds in the sample, the value increasing with decreasing molecular weight.[7] The SV of esters showed a 70% increase from that of the parent oil. This may be explained by the fact that the reaction, Equation 1, removes the glycerol molecule from the oil, thus reducing the average molecular weight of the resulting methyl ester.

3.1.3 Calorific Values

The gross calorific value enables an evaluation of the total heating energy generated during the combustion process. The gross calorific value of the methyl esters was found to be, on average, about 5 and 7% lower than that of kerosene and automotive diesel fuel respectively, as shown in Table 2.

The heating value depends on the composition of the fuel. Diesel No.2 and kerosene are comprised of paraffins, olefins and aromatics, a mixture that leads to high energy content per unit volume. On the other hand, JCME comprises long chain fatty acid esters of lower energy content. The higher density of the biodiesel fuel as compared to the diesel No.2 and kerosine compensates for the lower heating value of the biodiesel on a weight basis,[8] so the fuels from the two different origins have very similar heating values, volume for volume.

3.1.4 GC-MS Analysis

The ester compositions of the JCME are given in Table 3. Three main fatty acid esters were identified in the biodiesel fuel, namely the methyl esters of palmitic acid (17%), oleic acid (17%) and linoleic acid (43%). The biodiesel also showed an unknown peak with retention time of 79.5 minutes. Other components were as described in the literature, although the relative amounts differed.[3]

Table 3 *Composition of methyl esters derived from* Jatropha curcas

	Fatty acid	Retention time (minutes)	Composition (%)	Literature values (%)[3]
JCME	C16:0	79.677	17.27	15.1
	C16:1	78.677	5.4	-
	C18:0	86.952	3.6	7.3
	C18:1	86.427	16.8	34
	C18:2	85.927	19.8	
		86.177	23.14	43.3
	Unknown	79.477	13.94	-
	Total		99.94	99.7

3.2 Fuel Properties

The fuel parameters of JCME were determined at the Kenya Pipeline Company Limited using the ASTM procedures for testing diesels fuels.[9] The parameters obtained were compared with those of kerosene and diesel No.2 fuel, tested in the same laboratory using the same procedures. Results are shown in Table 4.

Table 4 *Comparison of fuel properties of JCME and two diesel fuels.*

Test	Description	Kerosene	Diesel No.2	JCME
	Appearance	Clear	Clear	Clear
D93	Flash Point (°C)	45.5	74	147.4
D1298	Density at 20 °C (g/ml)	0.7884	0.8480	0.8774
	Density at 15 °C (g/ml)	-	0.8621	0.8799
D86	Distillation (°C)			
	IBP*	157.2	192.0	268.5
	10%	170.0	239.0	315.0
	20%	176.0	263.0	329.0
	30%	181.5	278.5	333.0
	40%	187.0	290.0	335.0
	50%	194.0	301.0	337.0
	60%	201.0	313.0	341.5
	70%	209.0	326.0	344.5
	80%	219.0	340.0	344.5
	90%	231.0	358.0	349.5
	95%	240.0	371.0	361.5
	FBP*	261.0	385.0	368.5
	Recovery at 365°C (ml)	59.0	92.0	-
	Total volume (ml)	99.0	98.0	97.5
	Residue (ml)	1.0	1.0	-
	Loss (ml)	Nil	1.0	-
D445	Viscosity at 40 °C (cSt)	2.35	4.16	4.431
D95	Water content (%)	-	<0.05	0.06
D130	Copper Corrosion	-	1b	1b
D1500	Color ASTM	1a	1.4	0.4
D4738	Cetane Index	-	49.2	55.03
D482	Ash Content (%)	-	0.02	0.01

* *IBP – Initial Boiling Point; FBP – Final Boiling Point*

JCME has a flash point over 100 ^{0}C higher than that of kerosene, and over 70 ^{0}C higher than that of diesel No.2. This high flash point of 147 ^{0}C gives an indication of the absence of volatile compounds, such as residual methanol. It also complies with the ASTM Standard for biodiesel D6751, which requires that pure biodiesel has a minimum flash point of 130 ^{0}C.

The kinematic viscosity of *Jatropha* oil at 40 ^{0}C was 31.5 cSt. After transesterification to the *Jatropha* methyl esters, viscosity was reduced to 4.4 cST. The viscosities of kerosene and diesel No.2 were 2.35 cSt and 4.16 cSt respectively. Thus JCME has a viscosity 1.9 times that of kerosene and 1.06 that of diesel No.2. At 20 ^{0}C, the density of JCME was found to be 1.1 times that of kerosene and 1.03 times of that of diesel No.2.

The Calculated Cetane Index (CCI), though not an accurate predictor of Cetane Number for biodiesel, since it is based on a calculation using specific gravity and the distillation curve, was used to estimate the Cetane Number of the biodiesel and two diesel fuels. From the results it can be seen that biodiesel has a CCI value higher (55) than that of diesel No.2 (49). This accords the biodiesel with better cold-start properties, minimizes the formation of white smoke (emissions), and leads to less engine noise and hence improved engine durability and reduced fuel consumption.[9]

Distillation profiles of JCME (biodiesel), along with the two referenced diesel fuels (kerosene and No.2 diesels), are shown in Table 4. The data indicate that the biodiesel is less volatile than the diesel fuels. Kerosene was the most volatile with a final boiling point of 261 °C, which is less than the IBP of the JCME and about the boiling point of diesel No.2 at 20% recovery. Between 20 to 90% recovery, JCME recorded a narrower boiling range (1- 4°C) while the diesel fuels gave a range 5-12 °C for kerosene and 11-18 °C for No.2 diesel fuel. This is explained by the differences in composition of the fuels. Biodiesel from *Jatropha* constitutes mainly C18 fatty acid esters with closely similar boiling characteristics, giving the narrow boiling range. Kerosene and No.2 diesel, on the other hand, constitute a mix of paraffins, olefins and aromatics of different boiling points giving them a wide boiling range.[7]

JCME recorded a slightly higher water content than No.2 diesel fuel. On the other hand, the diesel No.2 gave double the ash content of JCME, mostly because diesel may contain sulfur and water-soluble metallic compounds as well as dirt and dust. JCME is sulfur-free, and without fossil residues.

4 CONCLUSIONS

In Kenya, the indigenous oil crop *Jatropha curcas* L. is abundant and can be easily cultivated. The plant is adapted to the prevailing harsh agro-ecological conditions in the arid and semi-arid areas which constitute over 70% of the country's land mass. The fuel properties of *Jatropha* biodiesel, such as density, viscosity and heating value, were found to be close to those of diesel No.2 fuel, and less close to those of kerosene. The alternative fuel can thus be used directly, or as blends with diesel fuel, in existing diesel engines, without any modifications. The higher flash point of the fuel adds the advantage of safety handling. This plant oil is thus a promising source of an alternative renewable and environmentally friendly fuel.

Acknowledgements

This study was supported by the Lake Victoria Research (vicRes) initiative. Mr. Festus K. Muchena, Chief Chemist of the Kenya Pipeline Company Ltd., facilitated the analysis of fuel properties at the company's laboratory. GC-MS analysis was carried out at the International Centre of Insect Physiology and Ecology, Nairobi. The Vanilla Jatropha Development Foundation supplied *Jatropha* seed. All are gratefully acknowledged.

References

1 GVEP (Global village energy partnership), Report for the Kenya sector energy need assessment GVEP workshop, Safari Park Hotel, Nairobi, 9-10th December, 2004; p 54.
2 X. Lang, A. Dalai, N. Bakhsh, M. Reaney and P. Hertz, *Bioresour. Technol.*, 2001, **80**, 53.
3 N. Foidl, G. Foidl, M. Sanchez, M. Mittebach Reaney, S. Hachel, *Bioresour. Technol.*, 1996, **58**, 77.
4 B. Furniss, J. Hannaford, G. Smith, R. Tatchell, *Vogel's Text Book of Practical Organic Chemistry*, John Wiley & Sons: New York; 1989; pp 400 - 401.
5 AOCS. *Calculated Iodine Value.* cd 1c-85: In: Official Methods and Recommended Practices of AOCS. American Oil Chemist's Society Press: Champaign; 1997b.
6 AOCS. *Acid Value.* Te-1a-64: In: Official Methods and Recommended Practices of AOCS. American Oil Chemist's Society Press: Champaign; 1997c.
7 G. Knothe, *J. Am. Oil. Chem. Soc.*, 2002, **79**, 847.
8 T. Russell, Additives influencing diesel fuel performance, John Wiley and Sons, Chichester, 1989, pp 65-104.
9 ASTM, D975-92a Standard Specification for Diesel Fuel Oils. Philadelphia Committee on Standards, 1992, pp 310-311.
10 V. Gerpen, B. Shanks and R. Pruszko, *Biodiesel Analytical Methods*, 2004, National Renewable Energy Laboratory, Iowa State University, Iowa.

JATROPHA BIODIESEL AS AN ALTERNATIVE TRANSPORT FUEL IN ZIMBABWE

C. S. Shonhiwa,[1] F. P. Gudyanga[2] and Z. Chiguvare[3]

[1]Renewable Energy Programme, Faculty of Engineering, University of Zimbabwe, Harare, Zimbabwe.
[2]Ministry of Science and Technology Development, 48 Samora Machel Ave., Harare, Zimbabwe.
[3]Finealt Engineering Private Limited, P.O. Box 2163, Harare, Zimbabwe.

1 INTRODUCTION

Following the oil crisis in 1973, the world saw the need to develop bio-fuels to reduce dependency on fossil fuels and hence improve the security of energy supply, especially for non-oil producing countries. This need is now more urgent still, due to the fact that world supplies of petroleum and natural gas are likely to run out within the next 42 and 63 years respectively if no new reserves are discovered.

Furthermore, 75% of these fossil-based resources are concentrated in a few OPEC countries and, under these circumstances, oil prices are more likely to remain unstable and are now hovering around US$ 123/barrel. It is estimated that the price of crude oil is likely to exceed US$ 170/barrel by the end of the year 2010. This situation cannot be sustained. The development of alternative energy resources will help stabilize oil prices and improve the security of energy supply for non-oil producing nations. The development and production of liquid bio-fuels, which has been ongoing since the oil crises of 1973, should contribute towards the attainment of this goal.

Countries that are at the forefront of the development of liquid bio-fuels are Brazil and certain countries in the European Union, especially Austria, Germany, Italy, France, Hungary and the Czech Republic (see Table 1). Today, Brazil produces about 50% of its own ecologically sustainable transport fuel, an achievement that has not yet been matched by any other country in the world. Development of liquid bio-fuels in the European Union was mainly driven by:

- The need to improve supply security of liquid fuel for the transport sector by having a renewable source at hand.
- The need to have an environmentally friendly fuel available for diesel combustion engines.
- The desire to reduce the health and safety risks associated with the use of petroleum products.
- The desire to provide customers with reliable fuel at reasonable cost-benefit ratio.

Country	Production capacity (tonnes/year)
Hungary	1 050 000
Germany	415 000
France	286 000
Italy	160 000
Belgium	85 000
Czech Republic	32 000
Austria	20 000
Sweden	6 000
United Kingdom	2 000

Table 1: *Biodiesel producers in Europe, year 2000*

Rapeseed oil methyl ester (RME) was the first biodiesel to be developed.[1] Biodiesels based on other feedstocks such as used cooking oil and sunflower oil are now also commercially available. The first pilot plant with a capacity of 500 tonnes/year was commissioned in 1988 in Austria. The first industrial scale biodiesel production plant went into operation in 1991. Its capacity was 10 000 tonnes/year. In the following years, biodiesel production plants of larger capacities were established all over Europe, e.g. the plant at Livorno in Italy has a capacity of 80 000 tonnes/year. Rouen in France has the world's largest biodiesel production plant with a capacity of 120 000 tonnes/year. Germany and Sweden also have biodiesel production plants of comparable capacities.

A study commissioned by the International Energy Agency and carried out by the Austrian Bio-fuels Institute in the late 1990s identified 21 countries around the world in which bio-fuel projects with a commercial objective have been implemented. The European Union still leads in research into biofuels and biodiesel production has also increased in the USA, with the most modern multi-feed-stock (MFS) biodiesel plant located in Kentucky.

Zimbabwe has been exposed to international sanctions, boycotts and embargoes by countries in Europe and North America. In response to this hostile environment, the Government has enunciated a "Look East Policy" that focuses on promotion of trade with more friendly nations like China, India and Malaysia among others. The current fuel crisis being experienced throughout the world (more so by Zimbabwe), Zimbabwe's political isolation, reduced industrial activity as a result of the recurrent droughts, reduced foreign currency generation and increases in global fuel prices have all contributed to the nation's inability to purchase the much-needed commodity. Thus, politically, the production of biodiesel on a national scale not only makes economic sense but also ensures that Zimbabwe is self-sufficient in the face of economic sanctions and changes in the international policies towards Zimbabwe. Strengthening the nation's capacity to produce biodiesel locally will cushion the economy from the adverse impacts of externally driven economic shocks.

Zimbabwe is dependent on an agro-based economy. As such, a number of oil plants that can be used in the production of biodiesels are grown in Zimbabwe. These include sunflower, groundnuts, soya bean, maize and jatropha. Oil from jatropha is the most suitable since it is not edible. The plant is grown as a hedge in some areas in Zimbabwe such as Mutoko, Binga, Victoria Falls and Nyanga. It can easily be grown in the arid and semi-arid agricultural zones of the country, where vast tracts of land lay fallow. It is not demanding in terms of management and can easily be grown at communal and commercial scale.

2 BACKGROUND TO JATROPHA

Jatropha originated in Central America and was introduced into Zimbabwe in 1995 by the Rockefeller Foundation (RF) and the German Government Technical Assistance Programme (GTZ).[2] It adapted well to Zimbabwean conditions, including semiarid marginal sites. The plant can withstand drought by shedding most of its leaves to reduce transpiration loss. The fallen leaves add to the organic matter in the soil and improve its fertility.

The plant can grow in a wide range of soils (even gravelly, sandy, saline and rocky soils) and climatic conditions (from high temperatures to light frost).[3] However, it cannot be grown on slopes exceeding 30^0, in highly alkaline soils (pH exceeding 9) or where the temperature falls below freezing point (which is a rare case in Zimbabwe). Furthermore, it cannot withstand extended water-logging.

It is a soft wooded shrub or small tree with whitish watery latex. Animals do not browse it. Seed yields vary from 2-3 kg/plant to 4-6 kg/plant. The oil content in seed varies from 25 to 48%. On average, one litre of oil can be obtained from 3.28 kg of seed.

Jatropha has a low gestation period (2 years) for seed production compared to other non-edible oil bearing plants. Some varieties start yielding from the first year onwards but the maximum production is reached in 3-5 years. The productive age of Jatropha is over 50 years. Based on soil and rainfall conditions, seed yields vary from 0.5 to 12 tonnes/hectare/year. In poor soils with low nutrient contents, a yield of about 0.5 – 0.75 tonnes of oil/hectare/year is expected.

2.1 Uses of Jatropha

A Steering Committee on development of jatropha biodiesel constituted by the Zimbabwean research scientists, engineers, economists and policy makers, recommended the cultivation of jatropha on account of several advantages to the economy. The benefits may be briefly summarised as follows:

1. Jatropha oil can be used as biodiesel and is highly compatible with petro-diesel up to 100% and can be used in diesel engines without having to make any change in them.
2. Glycerine, by-product of processing of jatropha oil for the production of biodiesel would reduce the cost of biodiesel and makes it economically viable since the operational costs would be abated by selling the glycerine which fetches more money (per unit cost) than the biodiesel produced.
3. Biodiesel would significantly reduce particulate and other emissions (such as carbon monoxide and hydrocarbons) from vehicles.
4. It would reduce the dependence of the country on imported fuel. Currently Zimbabwe imports all its fuel.
5. Jatropha can be grown on low fertility, marginal, degraded, fallow, waste and other lands (e.g., land along canals, highways or railways, on the borders of farmers' fields as a boundary fence) in different agro-climatic conditions, including semi-arid areas. It can also be planted around windmills in wind farms and around transmission towers and below power transmission lines. Jatropha planting reduces

soil erosion. It is not browsed on by animals.

6. Jatropha seed cake is used as an organic fertilizer and bio-pesticide. De-oiled seed cake has a crude protein content of between 53%–60% and a favourable amino acid profile. However, the seed cake is toxic due to phorbol esters, trypsin inhibitors, *etc*. On detoxification it can provide a highly nutritious feed material.
7. The cultivation of jatropha would generate employment on a large scale in rural areas almost round the year including the lean seasons. The oil may be extracted in mobile oil presses in a decentralized manner.
8. To meet the national demand of 1 billion liters per annum by 2020, as much as 265 000 hectares of land (which is about 0.65% of Zimbabwe's surface area) has to be brought under jatropha plantation.
9. According to the National Biodiesel Steering Committee set up in 2005, smaller units for the trans-esterification of jatropha oil into biodiesel were under trial and it would be possible to run the apparatus on a much larger quantity of feedstock.

One impediment to the cultivation of jatropha for the production of bio-fuel is that it would threaten the growing of food crops and the technology of commercial biodiesel production plant is not yet readily available on the market. However, the area of land that would be brought under jatropha cultivation during the initial years would not yield adequate quantities of seeds for a trans-esterification unit to operate economically. Farmers, on the other hand, would not go in for jatropha cultivation on a commercial basis without being assured of a reliable market. Clearly, it is necessary to find alternative uses for the oil in its natural form (i.e., without trans-esterification), for which a large quantity of the oil can be used for soap making.

2.2 Other Uses of Crude Jatropha Oil

People in many villages are familiar with the use of crude jatropha oil or other oils of plant origin (for example, castor oil) for the purposes of lighting and lubrication. The crude jatropha oil cannot travel up the wick in a lamp due to its high viscosity, and can thus be used only in very shallow lamps.

The Department of Mechanical Engineering - Renewable Energy Programme, University of Zimbabwe, has also studied the use of crude jatropha oil for burning in lamps for rural lighting called "marambi". Due to the oxygen content in the fuel and more complete combustion, soot formation of the oil is low compared to that of pure petrol-diesel. The chances of suffocation inside a house are also less while using jatropha oil for illumination purposes. Due to complete combustion, the luminosity of the flame given when burning a jatropha oil blend is comparable to that from conventional lamp oil (paraffin), although the size of flame is smaller. Due to its higher flash point, jatropha oil is safer in storage than lamp oil.

Also due to its high viscosity, it could not be used directly in the internal combustion engines of modern vehicles,[4] hence efforts were made by Eng. C.S. Shonhiwa of University of Zimbabwe to develop a biodiesel production unit which would change the physical properties of the crude jatropha oil to suit modern vehicles, and this was patented in 2004. The unit is able to convert crude jatropha oil from a highly viscous oil to an oil which is highly compatible with petrol-diesel.

3 CULTIVATION OF JATROPHA

Jatropha is propagated through seed and stem cuttings, with planting from February to March and September to October. Seeds must be collected from a genetically proven superior source for increased output. They need to be stored at a moisture content of 9-11% to prevent germination. The time limit for seed storage is 10 to 15 months. Seed testing procedures have been developed for maintaining quality control. Seed germination starts from the fifth day and takes up to 15 days. Germination percentage is 60 to 90%. Seedlings are grown in the nursery for three months and can be transplanted when 40 to 50 cm tall. Seedlings can be planted out after good rainfall. For propagation by cuttings, semi-hardwood cuttings of 15 to 30 cm length and 1 to 3 cm diameter may be planted in polythene bags. Rooting occurs in 8 to 15 days with 80% success. Cuttings can also be directly planted in the field.

Jatropha is commercially planted at a density of 1110 plants/ha, which corresponds to a 3 m x 3 m spacing. Planting pits should be of the size 30 cm x 30 cm x 30 cm. Weed control is needed during the early development and establishment phase. Manuring (fertilizer rich in nitrogen or organics) can be carried out during transplantation and after one year. One to two kg of cattle manure and 150 g of single super phosphate has been recommended per pit at the time of transplantation into the field, followed by 20 g urea after 30 days. Application of nitrogen and phosphorus promote flowering. Top pruning during March to May keeps the tree height down and makes seed collection easier.

The golden flea beetle (*Podagrica* spp.) can harm the young plants in particular. Jatropha is also host to the fungus "frogeye" (*Cercospora* spp.), common in tobacco. The flowers are pale yellowish-white and generally bloom between August to December. Fruits take two to four months to mature, when they turn yellow and are ready to be harvested. After picking, the seed coats are separated from the seeds by threshing or by using a de-shelling machine. The de-shelled seeds are sun-dried and used for oil extraction.

3.1 Jatropha Cultivation Schemes

The following five different schemes have been proposed for jatropha cultivation in Zimbabwe and are being implemented.

Scheme A

Cluster Village Scheme

One village would be chosen as the cluster village and this would be the central hub for all nearby villages. Volunteers would work in the base village first and would make this a pilot project. Later, the project would be extended to all the nearby villages. Initially, the work would start from plantation of trees (3 000 000+ seeds) in each cluster village. Seeds and cuttings would be donated to a responsible person in the village and also followed up by the cluster village coordinator.

The cluster village would be identified as the village where there were good contacts. This model scheme would then be perfected by running competitions in schools and communities, resulting in making it easy to convince nearby villages. The model would be made up of the following structures:

Village head – The person who lives in the village and is able to convince other farmers and communicate with our coordinators.

Village activist – In each village there would be a group of individuals, e.g. farmers or members of political parties, who were interested in doing something for the village.

Extended Village

Villages near to the cluster or model villages would be denoted as extended villages. Once the project was successfully implemented in the cluster village, it would be extended to nearby villages. The people from these villages would be brought to the model village and the advantages of the project would be explained to them. The seeds would be distributed again in the same fashion under the supervision of the village coordinator. Seeds produced would be brought to the cluster village for oil extraction. Thus one oil extraction unit could be utilized by all the villages in a district.

One active volunteer would be assigned to each cluster as the coordinator. The responsibility of the coordinator would be to oversee the entire project in his cluster and report to the district coordinator. All requests regarding funds and information would be directed to the district coordinator who administers the project in the whole district.

Goals of each cluster:

CLUSTER	
Jatropha Plantation	2 700 ha
Seeds Generated	20 000 tonnes/year
Oil production	6 000 tonnes/year

One Oil Extraction Unit (OEU) would be set up for each cluster, and would be responsible for processing the seeds collected from the cluster village as well as extended villages. Each cluster village would support the oil mill to exceed the maximum production capacity. Oil would then be transported to districts and then to a provincial biodiesel plant. Oil cake would be distributed to farmers to fulfill the local supply of organic fertilizer. Excess cake would be transported to districts and further to a Provincial Biodiesel Processing Plant for selling to other farmers.

A district would be made up of about 15 clusters. Its goals would be as follows.

Goals of each district:

DISTRICT	
Jatropha Plantation	40 500 ha
Seeds Generated	300 000 tonnes/year
Oil production	90 000 tonnes/year

One member would be chosen and assigned as a District Coordinator, with responsibility for administering the project in the district. Ten volunteers would be chosen and assigned as head of each cluster. 100 villagers who closely work in the project would be chosen and assigned as village heads. The 15 districts would feed the provincial biodiesel plant with at least 1 350 000 tonnes/year of jatropha oil.

Scheme B

Communal Farmers

These are small scale farmers found in every district with farm sizes ranging from 70 to 100 ha. These farmers would be encouraged to grow jatropha plants surrounding their farm land, with each farmer growing at least 2 ha. The estimated seed output from these farmers is 18 million tonnes/year.

Scheme C

Public Institutions

These include schools, prison farms, clinics, hospitals and universities. These institutions would be provided with inputs for growing jatropha around their premises and farms. It was estimated that each institution could provide an equivalent of at least 10 ha. The annual seed output is estimated at 600 000 tonnes.

Scheme D

Commercial farmers

These are large-scale farmers with vast areas of land which they use for crop production and animal ranching. These farmers could plant jatropha in their paddocks and each had a capacity to put at least 50 ha of land under jatropha cultivation. It was estimated that, nationwide, 5 000 farmers would be willing to take part in the scheme.

Scheme E

Plantations

The government had promised to provide about 5 000 ha of land per province to put under state jatropha plantations. These estates would be manned by the Forest Commission.

4 BIODIESEL PROCESSING PLANT

Biodiesel fuel can be produced from jatropha oil using a fairly simple chemical reaction known as trans-esterification. In this process, the oil reacts with a simple alcohol (methanol/ethanol) in the presence of a catalyst (caustic soda/potash) and, under specified conditions such as a temperature of 65°C and normal atmospheric pressure, yields a mixture of methyl or ethyl esters which is the biodiesel. This fuel can be used directly in diesel engines without engine modifications, and is quite comparable to petroleum diesel in its performance but with the added advantage of being environmentally friendly. Biodiesel burns more efficiently than mineral diesel because it contains its own oxygen (intrinsic oxygen) in its chemical structure. This means less atmospheric pollution. The reaction equation for biodiesel synthesis is shown below.

$$\textbf{Jatropha oil} + \textbf{alcohol (methanol)} \xrightarrow[\textbf{(1kg)}]{\textbf{Catalyst}} \textbf{biodiesel} + \textbf{glycerine}$$

(100 kg) (10 kg) (1kg) (100 kg) (10kg)

- Reaction temperature: 65°C
- Pressure: 1 bar
- Reaction time: 2 hours per batch.

The flow charts below (Figures 1 and 2) summarize the process of making biodiesel from the extraction of vegetable oil to production of the end product.

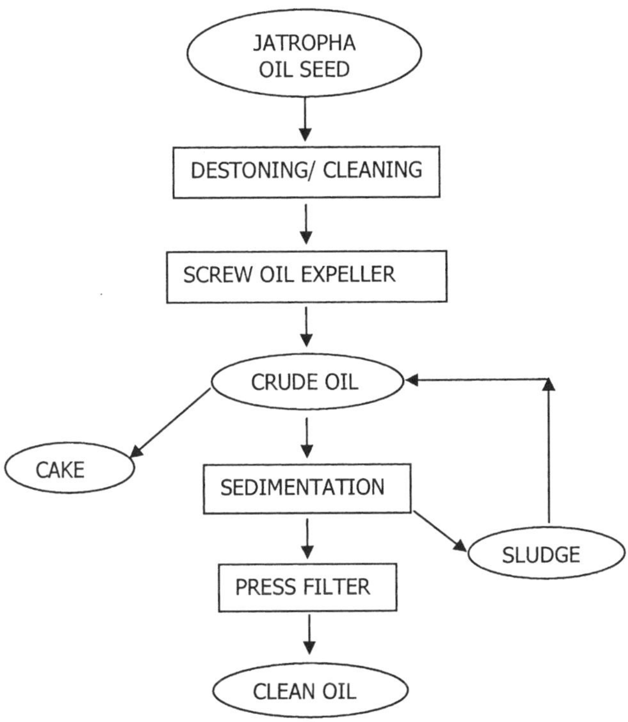

Figure 1: *Flow Chart for Biodiesel Production from Jatropha oil*

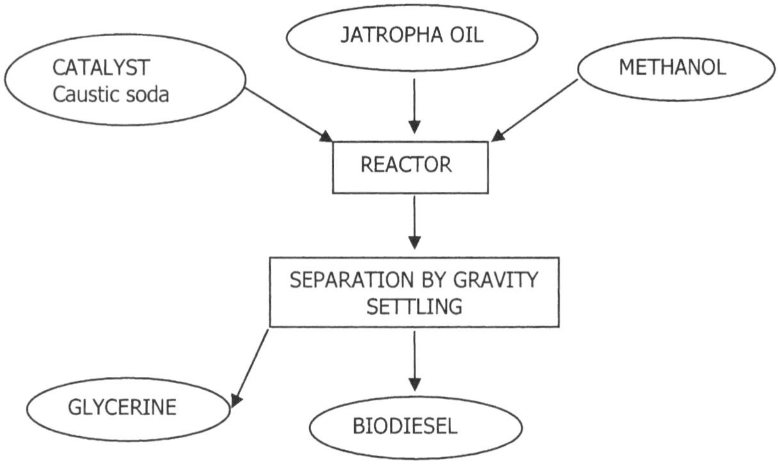

Figure 2: *Flow chart for oil production.*

4.1 Oil-expressing Machines

The small scale or farm-size oil expressing machines are intended for use by individual farmers or groups of farmers (cooperatives) who produce oilseeds such as jatropha, castor beans, sunflower and soya beans. The idea is to extract oil at the farms and then transport it to a biodiesel processing plant located at some centre in a district or province where it would then processed into biodiesel. The farm set-up will comprise an oil expressing machine and filter press to clean the oil.

4.1.1 The Design. The capacity of such machines can range from one to five tonnes of oil seed a day, which translates to 250 to 1250 liters of oil per day respectively. Machines in this range are already available on the market and there is no need to design and build others at the moment. Thanks to the 'Look East Policy' a filter press which goes with such machines is also available on the market from China. A typical oil expressing machine which can be used at Growth Points is a Chinese-made screw type machine with the following details:

- Type: - Model 6YL-100
- Capacity: - 1800 to 2400 kg oilseed per day.
- Price: - US$ 170 000
- Local supplier in Harare: Appropriate Technology Africa (along Seke Road)
- The filter press from the same supplier costs US$ 20 000.

Oil expressing machines would be distributed among individual farmers or groups of farmers throughout a district. The farmers choose the size of machines according to their affordability and needs. The oil from the farms is collected by the biodiesel processors using a model probably similar to the one used by the dairy industry when collecting milk from dairy farmers. This model may have the following advantages and disadvantages which need to be weighed against each other when deciding the best course of action.

4.1.2 Advantages

- Oilseed cake remains at the farm and can be used as stock feed (applicable to raw materials other than jatropha and castor bean) or as a fertilizer.
- Processed oil is less bulky to transport to the biodiesel plant than the unprocessed oilseeds.
- Extraction of the oil may be regarded as value addition to the product and hence should mean more income for the farmer.

4.1.3 Disadvantages

- Some residual oil remains in the cake because the expeller cannot exhaust all the oil, hence less oil yield per given mass of oilseed. Solvent extraction is neither economic nor safe at this level.
- Relatively high capital cost for acquiring an oil-expressing machine, filter press and solvent extraction technology.
- Handling of liquids is more problematic than is the case with solids.
- The process of oil extraction is time consuming for the farmers.
- Maintenance costs for the machines and filter press are prohibitive.
- The processed oil could encourage middlemen and side markets.

The national processing plant consists of two main sections, namely the Oil Extraction Plant and the Biodiesel Processing Plant.

4.2 Oil Extraction Section

The extraction section at the plant site will be composed of three modular oil extractions plants, each with the capacity to process 300 tonnes of jatropha seed per day. This translates to about 100 000 liters of jatropha oil per day (with an oil extraction efficiency of 30%) making a combined total of about 300 000 liters of jatropha oil per day (the density of jatropha oil being about 0.860 g/ml). The whole process may be divided as follows:

1. Raw material preparation
2. Extraction
3. Cake handling

4.2.1 Raw Material Preparation Stage. The main objective of the preparation stage is to ensure the seeds are in the best possible state for the later conditioning and oil extraction stages. At the cleaning stage, foreign materials like stones, leaves, trash and metallic objects are removed through use of a shaking table – screens, suction fans, cyclones and magnets. De-shelling is done to remove the coating (that is the hulls) from the seeds. Separation of the seeds from the chaff is done by winnowing and screening. The separated seeds are packed in jute bags and stacked in seed warehouses. The seeds would then be conveyed to the oil press machine for extracting the oil.

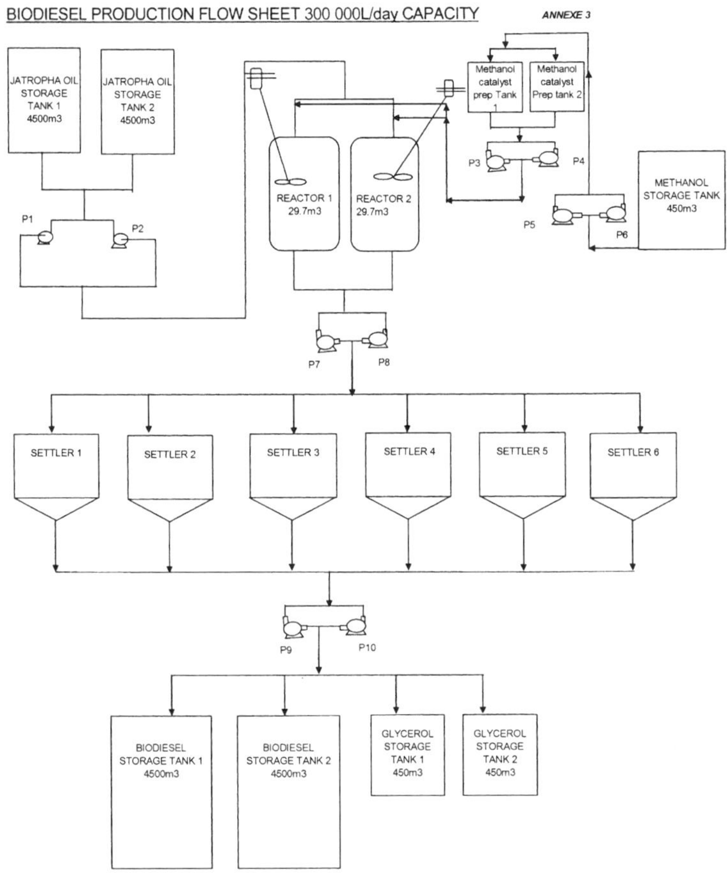

Figure 3: *Provincial Biodiesel Processing Plant*

4.2.2 Extraction. The dry seeds pass through the expelling stage where a screw press squeezes out the oil. The oil is screened and filtered, then pumped to the storage tanks. The cake residue is conveyed to the cake processing plant.

4.2.3 Jatropha Cake Processing. The jatropha cake is ground and then conveyed to the storage bins ready to be packaged where it will be used as an organic fertilizer.

4.3 Biodiesel Production Section

Jatropha oil collected from farmers or the oil extraction plant is stored in large tanks with a capacity of one month's supply of oil. Pumps P1 and P2 transfer the oil into six $5m^3$ capacity batch reactors each with an internal oil heating jacket. Each of the reactors has a simple explosion-proof paddle mixer, which has a fixed speed of 750 rpm. Methanol is added to the catalyst preparation tank using air operated pumps P3 and P4. The prepared alcohol catalyst solution is added to the reaction tank using pumps P5 and P6. Filling each of the reactors takes 30 minutes and the filling is done simultaneously. The reactants are agitated for two hours while steam for heating, generated by two boilers, is passed through the heating coils. Pumps P7 and P8 drive the reactants from the two reactors into a $60m^3$ cone-bottomed tank ready for biodiesel and glycerine separation. Emptying of each reactor takes 30 minutes and for the two reactors this occurs simultaneously. Each 8-hour shift uses two settling tanks, so a total of six of these tanks are required for the plant. The amount of biodiesel and glycerine is measured using a load cell. A Micro Motion coriolis-type density meter which is installed at the exit of the separation tank identifies the glycerine and biodiesel phases during the separation processes. A pump transfers the glycerine to the glycerine storage tank, after which the same pump then drives the biodiesel into the storage tanks that have the capacity to hold up to one month's production.

4.4 Regional biodiesel processing plant

This plant is made up of a single 300-tonne jatropha seed oil extraction unit and a 100 000-liter per day biodiesel production plant. The process is the same as in the national plant but the difference is only in capacity. These will be located at district or provincial centres.

5 PROMOTION OF BIODIESEL USE IN ZIMBABWE

Production of biodiesel for use in the transport sector is now well established in the developed world. Advanced production processes have been patented and are now commercially available on a turnkey basis. Zimbabwe's economy is agro-based and, as such, production of biodiesels would be a very viable option and will certainly improve the liquid fuel supply situation and safeguard the environment, especially in cities like Harare.

Initially biodiesel will be intended for agricultural machinery such as combine harvesters and tractors. Warranties have now been given for combine harvesters and tractors, e.g. SAME, Steyr, John Deere, Massey-Ferguson, Lindner and Mercedes Benz. With the development of sophisticated marketing strategies, the focus will be extended to other diesel-driven vehicles, such as buses in the public transport sector, and personal cars. For future purchases of diesel engines by public sector institutions such as the Zimbabwe Passenger Company (ZUPCO), District Development Fund (DDF), National Railways of Zimbabwe (NRZ), the Police Force, the Zimbabwe Defence Forces (ZDF) and Central Mechanical Department (CMED), warranties should be obtained for use of biodiesel from the respective manufacturers.

5.1 Cost-Benefit Analysis

A cost-benefit analysis of the project was carried out using foreign currency monetary system because of its stability. The currency that was used was the United States Dollar.

i. The calculations were based on the actual reported sales price in each case.
ii. We used a real discount rate (r) of 9.77%, calculated with the formula $r = [(1+i)/(1+p)]-1$; in which p is the inflation rate, assumed to be 4.4%, and i is the average interest rate on medium- and long-term loans, which was 14.6% in

2005.

iii. The calculations were made with an assumed yield of 8 tonnes of seeds per hectare per year (based on information from the interviews) and roughly 1110 trees per hectare.

iv. Seed output prices varied according to the project, from US$ 1000 per tonne paid by the government to US$ 150 per tonne paid by a local NGO.

Capital & Labor costs: buildings, machinery and equipment, labor as per respective activity. The cost of establishing a jatropha plantation has been estimated at US$ 723 per ha for the first three years. Table 2 shows the estimated capital costs, and operational costs for the production of 1800 tonnes, 5400 tonnes, and 7200 tonnes of jatropha biodiesel annually. The following cost heads have been considered in the analysis that gave the figures in the table:

i. Capital costs: Reactors, pumps, control panel, computer automation system, valves, mountings, methoxide tank, condensers, oil pressing machines, seed cleaners, seed elevators, blowers, conveyors, oil pumps, cyclones, fittings, labour, consultancy, induction course, and commissioning.

ii. Operational costs: Seed collection, depreciation of oil pressing plant, depreciation of biodiesel plant, cost of caustic soda and methanol, labour costs for oil expelling, labour costs for biodiesel production, soap making, glycerine purification, cake packaging, electricity and water costs.

	Biodiesel Plant size (tonnes of biodiesel per annum)	**Jatropha plantation required** (hectare)	**Cost of plantation establishment** (US$ for first 3 years)	**Capital costs for Plant establishment** (US$)	**Operational costs** (US$ per annum)
1	1800	824	596 000	148 000	1 345 000
2	5400	2472	1 787 000	278 000	4 029 000
3	7200	3296	2 383 000	577 000	5 371 000

Table 2 *Capital and operational costs for different sizes of Jatropha Biodiesel Industries*

When all has been considered, the average production cost of a litre of biodiesel was found to be US$ 0.75. If for instance the selling price is pegged at US$1 per litre, the total expected profit would be US$ 450 000, US$ 1 350 000 and US$ 3 600 000 per annum, respectively, for the plant sizes shown in Table 2. The payback period based on these profits and the original capital cost would be 1.65, 1.53, and 0.79 years, respectively. The expected lifetime of these plants is 20 years. The sale of the value added and other byproducts mentioned above can further reduce both the price of biodiesel and the payback period. It makes business sense to invest in jatropha agro-fuel production.

6 RESULTS OF THE RESEARCH

The results are based on information obtained from interviews and laboratory experiments carried out during the research period.

6.1 The data derived from the communities, both urban and rural, revealed that the project could be viable. Although the project was not initially well funded this project has a special significance. The nation is primarily interested in achieving energy self-sufficiency.

6.2 The major problem with growing jatropha as a cash crop relates to the lack of information that local villagers have on specific aspects of cultivation, together with their attitude towards market and price risk, which is partly culturally determined. However, it should be possible to overcome all these problems through training, demonstrations and a massive raising of public awareness.

6.3 It is not clear in which direction the jatropha chain will evolve. For example, it is not clear whether small oil-expelling units should be installed in village clusters, or whether a few large centralized ones should be built. Will the main technology be hand-presses or larger expellers? Currently, they are mostly working with small ram-press machines.

6.4 Another observation was that seed production is not yet sufficient to start running commercial biodiesel plants.

6.5 The physical and chemical properties of crude jatropha oil have been determined. It was observed that jatropha oil is dominated by stearic, oleic, linoleic and linolenic acids. Most of these acids are unsaturated hydrocarbons, i.e. have double bonds, and therefore can be oxidized by air. Oxidation is not desirable, so both the crude oil and the biodiesel have to be protected by a nitrogen blanket when stored. If the storage vessel is air-tight, oxygen from the enclosed air may react with the diesel or oil, hence creating a partial vacuum. This may lead to implosion of the storage vessel if not appropriately designed.

6.6 As far as the utilization of the seed cake is concerned, the research revealed that the seedcake could pressed into briquettes, which can be used as fuel in wood stoves or ovens. Using seedcake as fertilizer could be more promising because of its favourable nutritional qualities. This possibility was mentioned by several respondents, but it remains untested as yet. Potential actors in this domain are farmers who want to use the cake as fertilizer, and the oil pressing facilities, which generate jatropha cake as a by-product.

6.7 The use of jatropha oil as fuel in oil lamps was another interesting use due to the scarcity of paraffin within the country. Some individuals have embarked on producing jatropha oil lamps, but there are some scientific aspects that need to be addressed to improve the efficiency of the lamps, particularly with respect to the wick, which is too thick.

6.8 Small-scale jatropha soap making has been undertaken for several years by local women's groups assisted by an NGO. Even though this application is not energy related, its value lies in the many learning experiences it has generated for local women and their NGO partner with the cultivation, pressing and use of jatropha. It has also contributed to network formation. The soap itself is a good product with strong antiseptic qualities.

6.9 Engine test runs were carried out on diesel engine with various diesel/jatropha diesel blends and the following results were obtained. The extent to which carbon had been deposited on the injector jets, piston and valves was noted after the engine had been run for 100 hours. The results are shown in Figures 4, 5 and 6, below. It was concluded that there was no significant carbon deposits on the injectors, pistons and valves, which shows that there was complete combustion of the fuel.

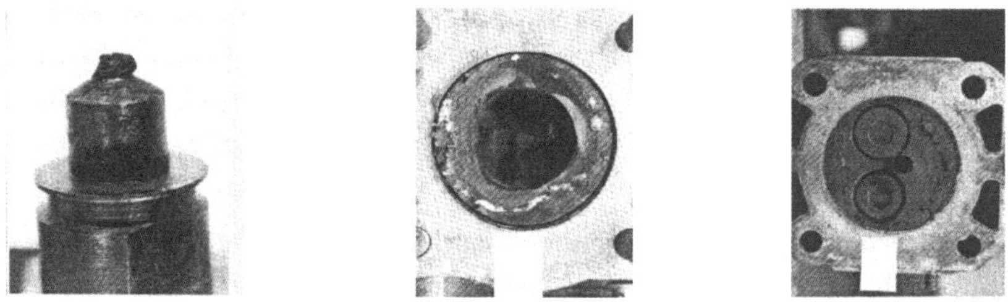

Injector jet **Piston** **Valve**

Figure 4 *Engine Test Run with 100% Diesel*

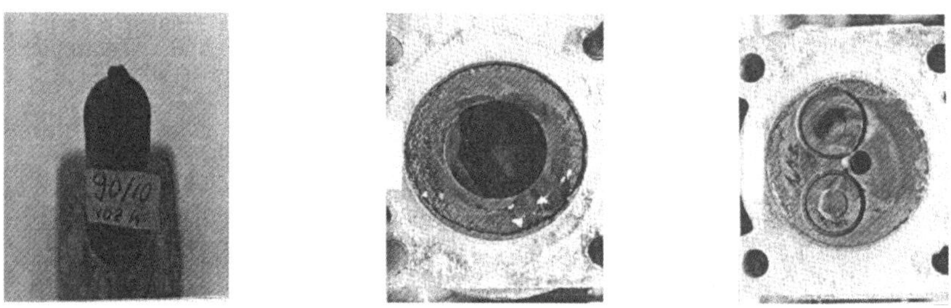

Injector jet **Piston** **Valve**

Figure 5 *Engine Test Run with 90% Diesel and 10% Jatropha Biodiesel*

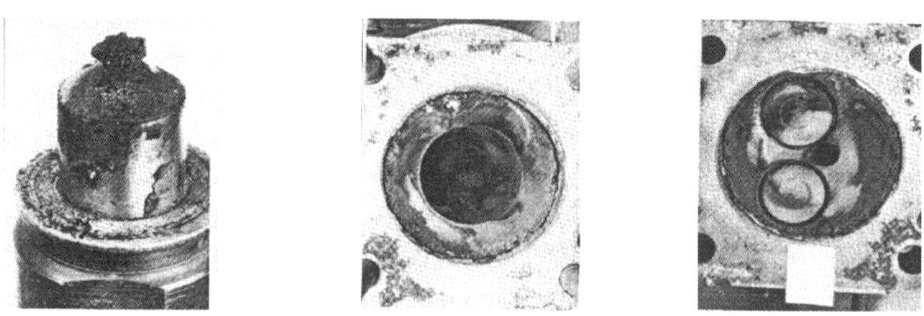

Injector jet **Piston** **Valve**

Figure 6 *Engine Test Run on 100% Jatropha Biodiesel*

7 CONCLUSIONS

Several economic gains arise out of the cultivation of jatropha. It will bring into use vast areas of wasteland in the State, prevent soil erosion and improve the quality of the soil. It will generate rural employment on a large scale. The produce in the form of straight jatropha oil may be used for lighting, lubrication and cooking as well as for making of soaps and candles etc. Once an adequate area of land comes under jatropha cultivation, biodiesel units can operate profitably and may be set up. An increasing use of domestically produced biodiesel would contribute greatly to relieving Zimbabwe's dependence on imported oil. There would also be environmental and health benefits from the lower emission of pollutants from equipment using biodiesel.

References

1 S. Saka and D. Kusdiana, "Biodiesel Fuel from Rape Seed Oil as Prepared in Supercritical Methanol", *Fuel*, 2001, **80**, 225.

2 Plant oil: Zimbabwe's sustainable fuel for the future, *Zimbabwe Biomass News*, 1996, Vol. 1, No. 2, BUN-Zimbabwe.

3 J. Heller, 1996. Physic Nut Jatropha curcas L. Promoting the conservation and use of underutilized vegetable crops: Institute of plant genetics and crop plant research.

4 Engine Manufacturers' Association (EMA). Biodiesel fuels and their use in diesel engines. Chicago, IL, August 1995.

BIOMASS, BIO-BASED PRODUCTS AND BIOENERGY

P.C. Vieira

Departamento de Química, Universidade Federal de São Carlos, São Carlos, SP, Brazil.

1 INTRODUCTION

The use of biomass, bio-based products and bioenergy requires an interaction between man and nature. A search on the internet using the keywords "human interaction with nature" led to the site of the Dryden and District Museum, Ontario, Canada.[1] Interestingly, the museum has an exhibition showing the intervention of man in that region and specially calls the attention of young people to the necessity of the new generations to think of how man should interact with nature in a sustainable way. So we can make this statement: "Human interaction with nature should explore the relationship between humans and the natural environment as a defining factor in the development and sustainability of a community".

Continuing the search it was possible to hit the site "Plants for a future", where a variety of uses of plants is mentioned.[2] It highlighted the use of plants as food, as medicines and for many other purposes like lubricants, fuel, for lighting, in paints and varnishes, etc. Here we will focus our discussion mainly on the possibility of using plants as sources of bioproducts and for biomass to be transformed into biofuels.

Nature has afforded a number of products to man, and among them are the so-called bioproducts which can be defined as sustainable, new, environmentally friendly products, or products for new applications that are derived from renewable sources, and based on advanced technologies and eco-efficient processes. Examples of these bioproducts are biofuels, bioenergy, biomaterials, and a vast spectrum of chemicals and biochemicals such as nutraceuticals and those connected with biotechnology such as genomics and bioprospection. Bioproducts come from the forest, from agriculture, from the sea, and from other more surprising sources such as garbage.

2 BIOPRODUCTS AND THEIR ORIGIN FROM BIORENEWABLE RESOURCES

Among the commercial bioproducts are biomaterials, fuels and chemicals. Some of the most common biomaterials include soybean oils, inks, pigments and dyes, paints and varnishes, soaps and detergents, industrial adhesives, biopolymers and films, and composite materials.

2.1 Fuels

Fuels can be divided into three categories: solid, liquid and gaseous. Here we are going to discuss mainly the use of biomass to afford liquid fuels. However, biomass is also an important source of solid and gaseous fuels. Examples of solid fuels obtained from biomass are coke, lignin, and bagasse. Examples of liquids are fuel oil, ethanol, and methanol. And examples of gases are syngas, methane, and hydrogen.

2.2 Chemicals

Several different types of chemicals can be derived from natural sources. Some of the more popular commercialized examples are activated carbon, oxy-fuel additives, phenols and furfural, specialty chemicals, acetic and fatty acids, industrial surfactants, and agricultural chemicals.

The use of bioproducts is attractive for a number of reasons. They allow the use of currently unexploited productivity in agriculture and forestry; they allow reliance on products that are more biodegradable and processes that create less pollution and generally have fewer harmful environmental impacts; they lead to the development of less expensive and better-performing products, and of novel materials not available from petroleum sources; they permit the exploitation of capacities in the field of molecular biology to selectively modify raw materials and reduce the costs associated with the production of raw materials and their processing; they lead to the revitalization of rural economies by the production and processing of renewable resources in smaller communities; they reduce the potential for the disruption of an economy due to dependence on imported fuel; they help to counter oligopoly pricing on world petroleum markets; and finally, they assist in the mitigation of projected global climate change through a reduction in the build-up of atmospheric carbon dioxide.

Brazil has been associated with the use of bioproducts since its discovery. One example is the exportation of brazilwood, which started in the sixteenth century and continues to this day. This wood is valued for its resistance, but it also produces a pigment that has been used by craftsmen for many years as an important source of a variety of colors (Figure 1). Through acid treatment with vinegar or yeast, the craftsmen could obtain different shades of red and, when made alkaline with ash, other colors such as purple could be obtained.[3] Pigments from plants have also been used by the native Indians, a good example of which is the red carotenoid pigment known as urucum.

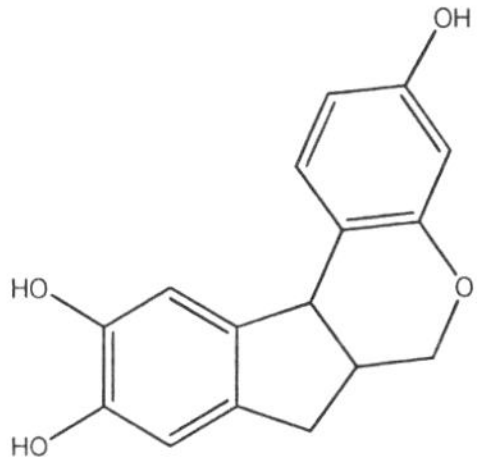

1

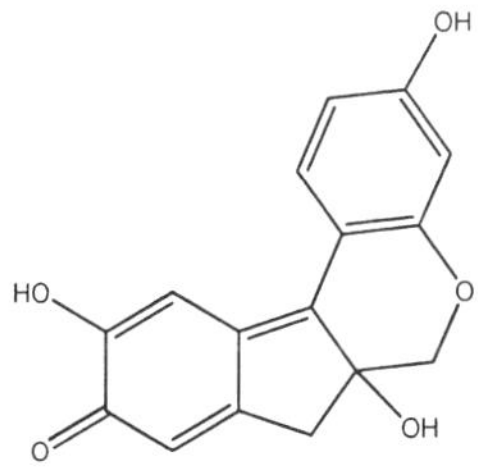

2

Figure 1 *Structures of brazilin* (**1**) *and brazilein* (**2**) *found in the pigments of brazilwood tree*

Plants have afforded many different compounds, and thinking particularly of pigments, it is worth mentioning an initiative by the American Chemical Society who in 1999 organised a program called World of Color, featuring the use of natural plant materials in a global experiment to map natural sources for eight different colors. The Brazilian Chemical Society (SBQ) participated in the program by launching a competition in which high school students painted pictures with plant pigments. Using pigments obtained from "jenipapo", "ipê", eucalyptus, mauve, "jaca", guava, coconut, cashew, cabbage, beet, "açaí", grape, annatto, safflower, coffee, etc., these students could express their feelings in a very beautiful example of how to connect chemistry and the arts.

Other important examples of bioproducts from plants relate to their use as medicines. Plants have long been known to afford secondary metabolites that are useful in the treatment of many diseases. One good example takes us back to the beginning of civilization in South America when the Spanish navigators began the colonization of the Amazon basin and discovered that the native Indians were using the bark of a plant known as cinchona to treat fevers. It emerged that the fevers were the symptoms of malaria, and the compound responsible for the cure was quinine, which controlled the causative agent, the protozoan parasite *Plasmodium falciparum*. The chemical structure of quinine was later determined by two French investigators, Caventou and Pelletier (Figure 2). Over a long period, quinine was the only product available for the treatment of malaria. However, with the advent of resistance, other chemicals had to be found and the synthetic compound chloroquine was developed to solve the problem of quinine-resistant strains. More recently, and again due to resistance, this time to chloroquine, new antimalarials had to be found. Once again, the inspiration came from plants, this time from traditional Chinese medicine. The novel sesquiterpene artemisinin, isolated from a plant used by the Chinese for the treatment of malaria, proved to be effective against chloroquine-resistant strains of *P. falciparum* (Figure 2). Not only artemisinin but also some of its derivatives have been used successfully in the treatment of malaria. So quinine and artemisinin are two examples of the importance of bioproducts as medicines.

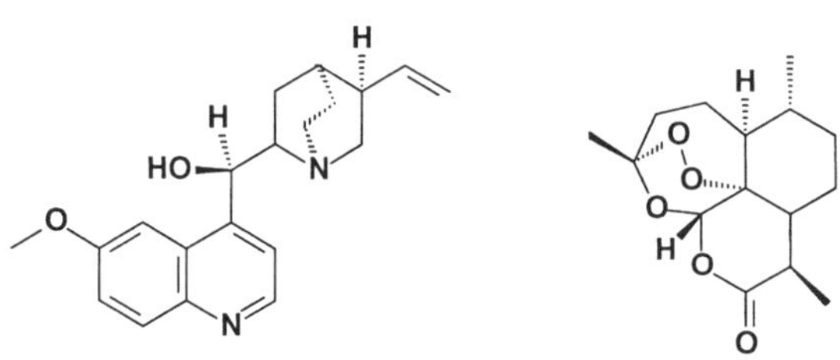

3 4

Figure 2 *Structures of quinine* (**3**) *and artemisinin* (**4**)

Projects relating to the use of bioproducts have been conducted in our laboratory at the Federal University of São Carlos in Brazil. Three main projects can be mentioned. The first involves studies on the potential of certain plant species to produce natural products for the control of leaf-cutting ants. The second is on the theme of phytochemistry and chemical ecology, and concerns the search for model compounds for the development of new insecticides, fungicides and bactericides to control plant pests. And the third concerns a search for antiparasitic natural products.

3 BIOMASS AND BIOENERGY

The use of biomass as a source of energy has long been known. Since ancient times, wood has been used for the purposes of heating and lighting. More recently, other sources of energy have emerged. For example, animal fats had an important role as fuels for lighting for many years. Petroleum and natural gases have now replaced most other sources of energy because of their abundance, low prices, and convenience. However, the widespread and indiscriminate use of fossil fuels has led to problems of pollution and global warming due to the emission of carbon dioxide and other gases. With these problems in mind, it is clear that alternatives to fossil fuels are essential.

In contrast to many other countries, Brazil obtains a large proportion of its energy, up to 46%, from renewable sources, of which biomass and hydroelectricity are the most important (Figure 3).[4] A comparison of these figures with those from the countries of the European Community and the rest of the world shows that Brazil is far ahead in its use of renewable sources for its energy requirements.

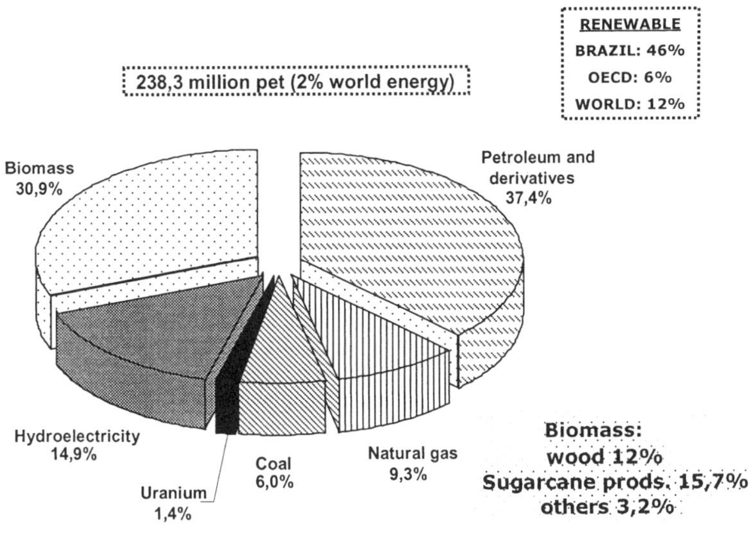

Figure 3 *Sources of energy in Brazil, 2007 (%)*[4]

Biofuels, which are liquids, are useful alternatives to fossil fuels. Their use as replacements for gasoline and diesel is attractive for a variety of reasons, mainly because global petroleum reserves are limited and new sources are located in geographic areas of conflict or where exploration will be costly, and because of the consequences for climate change that follow from the burning of fossil fuels. In this context, for a biofuel to be viable, it needs to provide a net energy gain, to have environmental benefits, and to be economically competitive. It must also be possible to produce it in large quantities without compromising food supplies. Biofuels can only be considered to be truly sustainable if their use stabilizes and reverses global climate change, minimizes the political and economic volatility associated with energy production and distribution, and assists the transition from fossil fuels into the future.

Biofuels ultimately derive their stored chemical energy from the sun *via* photosynthesis. Solar radiation is clearly a sustainable source of energy on the human timescale. However, the reactions of photosynthesis impose a maximal efficiency of only 12% and, in practice, final yields are significantly lower than this. So, in some ways the use of biomass to provide energy is not favorable. However, considering the points raised above, it does seem reasonable to explore the use of biofuels as alternatives to petroleum-based fuels that should be less harmful to the environment.

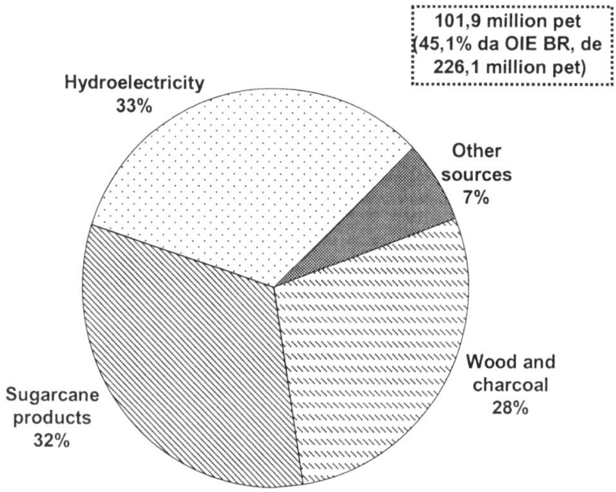

Figure 4 *Renewable energy matrix in Brazil, 2006 (%)*[4]

So, in the next section, we will focus on the role that Brazil is playing in the development of its program on ethanol and biodiesel. We will also briefly discuss the importance of some processes in the production of hydrogen, and of derivatives of 2,5-dimethylfuran that seem to be promising for the use as biofuels and are also derived from sugar.

The development of new biofuels needs to take into account their heating power in comparison to that of gasoline, the ways in which they can be obtained through the processes of biological engineering, and their compatibility with the existing infrastructure.

A quick look at the most promising biofuels leads to the conclusion that ethanol and butanol are those where the use of biotechnology could play an important role. The same cannot be said for hydrogen and biodiesel, since these cannot be obtained directly by biotechnological processes, nor for 2,5-dimethylfuran, since chemistry is vital for the development of its derivatives.

3.1 Ethanol

Ethanol is the most successful biofuel and, for example, already supplies 40% of Brazil's transportation fuel needs. Its main advantage is that there is a well-established infrastructure for its production and use. Techniques for its production by the fermentation of sucrose are in place, large-scale distillation technologies have been developed, and it can be used in so-called flexible-fuel vehicles. In addition, the logistics for its distribution are well-established. Every gas station in Brazil supplies it, with gasoline being provided in a commercial blend containing up to 24% ethanol.

However, several disadvantages have been attributed to ethanol. For example, it can be harmfully corrosive to the engines of vehicles, though this is not a problem in Brazil. In addition, its miscibility with water increases its price because distillation is required for obtaining both the hydrated and then the dehydrated form of ethanol. The lower power of ethanol compared to gasoline is another disadvantage that should be considered, ethanol containing around 60% of the heating power of gasoline. It is also important to mention that the process of fermentation very much depends on the availability of sugars like sucrose and fructose, and this restricts the types of plants that can be used to provide the required biomass. Much effort has been made to use cellulose as a starting material for the production of sugars for conversion into ethanol by fermentation, but this technology is in its infancy, and is not commercially viable so far.

In Brazil, ethanol is produced exclusively from sugar cane. The whole biomass of the sugar cane is used in the production of energy, and the figures below give an idea of just how efficient this process is. After crushing the cane, the juice can either be transformed into sucrose or ethanol. If the intention is the generation of energy, the juice is fermented to yield a mixture of water and ethanol from which ethanol can be obtained by distillation. Alternatively, the juice can be transformed into sucrose that generates molasses as a by-product that can then be fermented to produce ethanol. The bagasse can also be used as a source of energy in a process called co-generation. It is burnt to provide heating for the distillation of ethanol, with excess heat being transformed into electricity to be used in the sugar plant and, where there is a surplus, to be sold to the public distribution system. It is expected that when a method for the transformation of cellulose into ethanol has been perfected, the bagasse and leaves of the sugar cane will provide further ethanol and a further improvement in the yield of energy.

During the first world oil crisis back in the 1970s, the Brazilian government launched a program called pro-alcohol. Initially, this program was criticized and not many people, even the scientists, believed that it would be successful. However, 30 years later it is clear that ethanol from sugar cane is the most successful biofuel in world and already supplies a large percentage of Brazil's transportation fuel needs.

From 2003 to 2008, the production of ethanol in Brazil increased from 14.5 billion to approximately 21 billion liters. It is expected that by the year 2010 Brazil will be producing more 23 billion liters of ethanol. The consumption of ethanol has also been rising in the past few years, especially due to the fact that the large majority of cars running in the country today can use either ethanol or gasoline as fuels. Since prices favor the use of ethanol, most drivers prefer ethanol in the tanks of their cars. In February 2008, for the first time, the consumption of ethanol (1.432 billion liters) was larger than that of gasoline (1.411 billion liters) and the consumption of ethanol is expected to continue to grow in the future.

Looking at the regions that produce ethanol in Brazil, it is clear that the south-east is far and away the largest producer, being responsible for about 70% of the national production, followed by the center-west at 13% and the north-east at 9%. The northern region, including the Amazon basin, accounts for only 0.4% of the country's production.

The production of high quantities of ethanol in Brazil is the result of several factors, such as the development of improved strains of sugar cane and better fermentative processes, and better utilization of the land where sugar cane is grown. Taken together, these factors create highly favorable conditions for the efficient production of ethanol. This has led the chemical industry to look at ethanol as a starting material for the production of what has been called green plastic. Ethanol can be transformed into ethylene, the starting unit for the production of plastic.

Already the eighth-largest producer of petro-based plastics, Brazil will soon be the largest producer of organic plastics, according to Dow and Braskem. Both companies report that they have mastered the technologies that allow them to turn sugar cane into polyethylene, the most popular plastic. By 2012, about 10% of Brazil's plastic will come from cane instead of petroleum. Dow and Braskem plan to produce 1.2 billion pounds of organic plastic in Brazil by 2012. Of course, the economic production of green plastic depends on the price of petroleum, and it has been said that plastic obtained from ethanol will only be economically viable if the price of petroleum is higher than US$ 40 a barrel. The price of a barrel of petroleum recently rose above US$ 150, creating favorable expectations for the production of green plastic. However, the current world economic crisis has pushed the price of petroleum down again, and this, in the short term, may compromise the production of plastic from ethanol, even though it is also always important to consider that green plastic is more environmentally friendly than that obtained from petroleum.

3.2 Biodiesel

The following definition of biodiesel for Brazil was adopted in September 2005, law n° 11.097, which introduced biodiesel to the Brazilian energy matrix: "*Biodiesel: biofuel derived from renewable biomass for use in engines of internal combustion by ignition for compression or, according to regulation, for the generation of a different type of energy to replace partially or completely the fuels from fossil origin*".

Some countries have already adopted biodiesel as a fuel for cars and trucks, it being blended with diesel in different ratios. For example, diesel bought in Brazil today contains 5% biodiesel. Clearly, Brazil's requirement for biodiesel is large, and production is predicted to reach 3.3 million liters by the year 2010.

Biodiesel is made up of the methyl or ethyl esters of fatty acids, and is obtained from the triglycerides found in vegetable oils through the well-known transesterification reaction (Scheme 1).

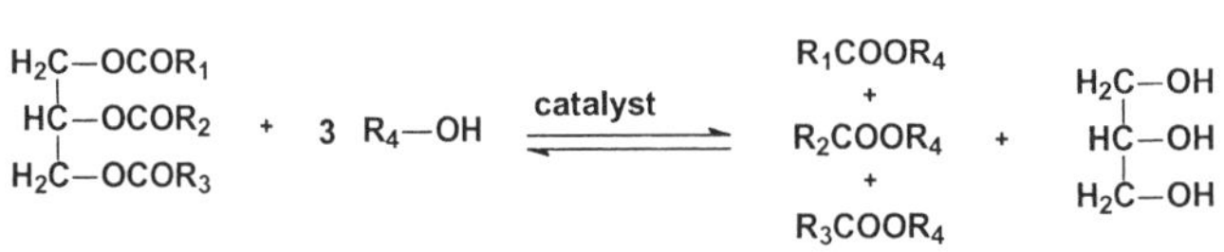

Scheme 1 *General transesterification reaction of triacylglycerides*

One of the problems associated with the production of biodiesel is its quality, especially its contamination with by-products. As the transesterification occurs in several steps, one of the main problems is the contamination of the final product with various intermediates of the process, as well as with the by-product glycerol (Figures 5 and 6).

The production of biodiesel is based mainly on plant oils as starting materials (although animal fats may also be used). Table 1 shows the most important oily plants and the yields of biodiesel obtained from them. There are high expectations for the use of jatropha, which can afford a very high yield of biodiesel, and dende is also a very promising oleaginous plant. Programs of massive plantation of some of these plants have been going on in several countries.

The production of biodiesel in Brazil is highest in the north-east of the country, that region accounting for around 50% of the total, followed by the south-east, which produces around 30%. The north of Brazil, including the Amazon basin, produces less than 4% of Brazil's biodiesel.

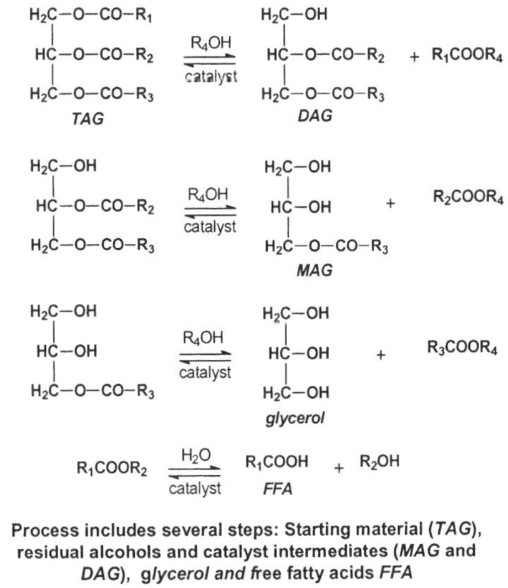

Figure 5 *Process of production of biodiesel by transesterification*

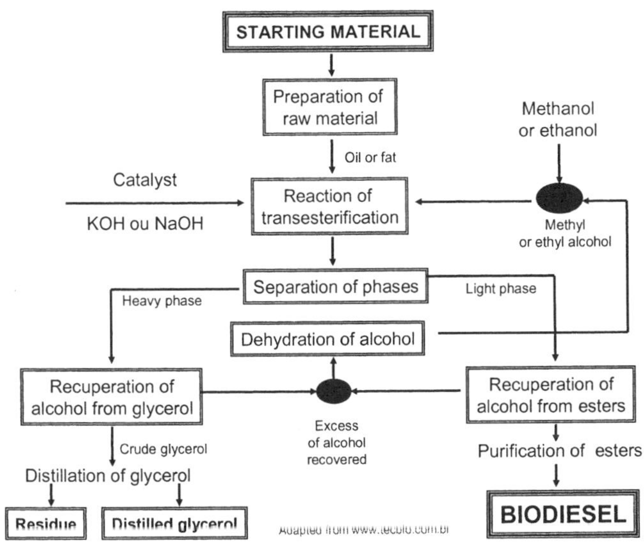

Figure 6 *Fluxogram representing the production of biodiesel*[5]

3.3 Hydrogen

Hydrogen has many advantages over alternative fuels. It is non-polluting and non-toxic; it is easily stored and transported; and, when produced by fermentation, it is easy to separate from the culture medium. On the other hand, it has certain disadvantages. For example, there is no infrastructure to store and transport hydrogen, and there are few vehicles that can run on it. In addition, the hydrogenases that produce hydrogen have bimetallic active sites and are extremely sensitive to oxygen. For these reasons, hydrogen will not be included in the Brazilian energy matrix in the near future (Figure 7). It is also interesting to mention that, if introduced, hydrogen will initially be produced by reforming natural gases, then by the electrolysis of water and reforming ethanol, and only by 2025 could it possibly be produced by alternative methods, perhaps by biotechnology.

Table 1 *Some of the plants with potential to be used as a source of energy*

Species	Productivity (tons/ha)	Percentage of oil	Life cycle	Regions and states*	Type of crop	Yield (tons oil/ha)
Cotton	0.86-1.4	15	Annual	MT, GO, MS, BA, and MA	Mechanized	0.1-0.2
Peanuts	1.5-2	40-43	Annual	SP	Mechanized	0.6-0.8
Dende oil	15-25	20	Perennial	BA and PA	Non mechanized	3-6
Sunflower	1.5-2	28-48	Annual	GO, MS, SP, RS and PR	Mechanized	0.5-0.9
Castor beans	0.5-1.5	43-45	Annual	Northeast	Non mechanized	0.5-0.9
Jatropha oil	2-12	50-52	Perennial	Northeast and MG	Non mechanized	1-6
Soya beans	2-3	17	Annual	MT, PR, RS, GO, MS, MG and SP	Mechanized	0.2-0.4

*MT: Mato Grosso, GO: Goiás, MS: Mato Grosso do Sul, BA; Bahia, MA: Maranhão, SP: São Paulo, PR: Paraná, RS: Rio Grande do Sul, MG: Minas Gerais

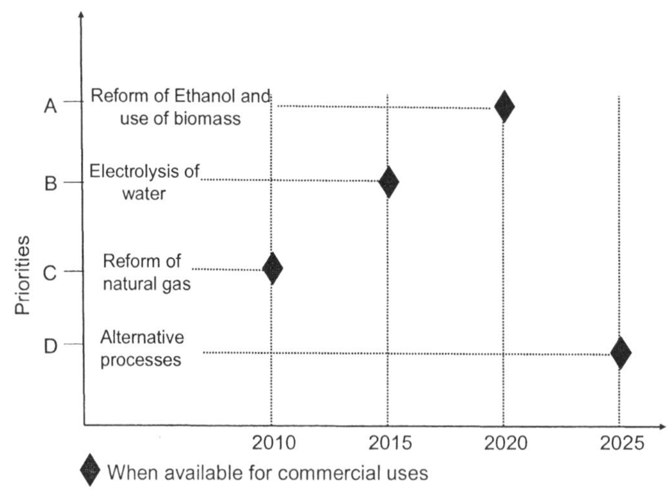

Figure 7 *Sources of hydrogen for introduction to the energy matrix in Brazil*[4]

3.4 Biomass

Biomass comprises cellulose (40–50%), hemicellulose (25–35%) and lignin (15–20%). Its use in fermentation processes requires a pre-treatment that includes milling to reduce the size of particles, thermochemical treatment to remove lignocellulose, enzymatic processing to release sugars and, finally, fermentation for the production of ethanol. There are high expectations for this process, and a number of investigators are working in this field to make it more efficient, so that it can be used in a way that is economically and energetically favorable. Once this goal is achieved, it will be possible to transform almost any biomass into biofuel.

The authors of a recent paper in *Nature Biotechnology* state that, for cellulosic ethanol to become a reality, research should be focused on finding biotechnological solutions that optimize the conversion of biomass into sugars, and this is certainly one area that deserves attention.[6] In another issue of the same journal,[7] biotechnology is again claimed to hold an important role in providing methods for the conversion of biomass into biofuels, this time by focusing on ways in which metabolic engineering can produce the next generation of biofuels. Certainly, the use of biotechnology to control metabolic processes so that genes and enzymes produce the desired products holds great promise.

4 OTHER POSSIBILITIES

Butanol is another possible liquid biofuel, and it has several advantages over ethanol, e.g. it has a longer carbon chain, resulting in a higher energy content, its extraction is more straightforward due to its insolubility in water, and it has a lower vapor pressure and is less corrosive. On the other hand, most biological systems are not adapted to produce butanol, so the ways in which it can be obtained through microbiological processes is very limited. However, the organism *Clostridium acetobutylicum* is known to ferment sugars to produce acetone, butanol and ethanol (the ABE process). This is the starting point for genetic engineering to provide organisms capable of producing higher proportions of butanol, highlighting once more the importance of biotechnology in the production of alternative liquid fuels.

Dimethylfurans are also potential fuels of the future, and a number of investigators are looking at chemical methods for the transformation of sugars into these compounds and their derivatives. So, starting with glucose or fructose it is possible to access dimethylfuran derivatives in a series of chemically-controlled reactions.[8,9]

5 PERSPECTIVES

A recently published paper provides a good and detailed discussion on the prospects for biofuels, and mentions a number of issues relating to the use of this alternative form of energy.[10] A paper entitled *New World For Biofuels* appeared in a recent issue of *Chemical and Engineering News.*[11] In it, Brent Erickson, the Vice President of the Biotechnology Industry Organization, stated that: "Ethanol production standards will transform biofuels business". He also claimed that: "We are at a point where we were in the 1850s when kerosene was first distilled and began to replace whale oil. This technology will be coming so fast that what we say today won't be true in two years".

A new U.S. law requires that 36 billion gallons of ethanol be produced per annum by 2022, of which 15 billion will be from corn kernels and 21 billion from cellulosic feedstock. Currently, U.S. farmers produce 7 billion gallons of corn ethanol, and there are no commercial plants able to refine ethanol from cellulose feedstocks. The industry experts predict that some 300 new refineries will be built, 210 to process cellulosic feedstock and 75 to be based on corn. Erickson compared the scale of ethanol-related economic impacts from the new law to those of the Manhattan Project.

5.1 Myths about ethanol

It has been suggested that ethanol...

- ...is responsible for deforestation, but this is not true, since sugar cane occupies only 2% of the 2.9 million Km^2 of the agribusiness sector in Brazil.
- ...is responsible for reducing the supply of food. However, Brazil and the USA together use only 0.1 million Km^2 for the production of biofuel out of the 12 million Km^2 used globally for agriculture.
- ...produces no emission of CO_2. This is not true, since the production of ethanol requires energy inputs, but the key point is that the input of energy is always lower than the energy that can be derived from the resulting ethanol. It is much more efficient to produce ethanol from sugar cane than from corn. In the production of ethanol from corn, each unit of energy (U.E.) used to produce it leads to only 1.5 U.E. coming out, whereas in the production of ethanol from sugar cane, for each U.E. used to produce it, about 8.0 U.E. can be obtained from the ethanol produced.
- ... is viable only in a limited number of places, like Brazil. However, sugar cane is grown around the world, of which Brazil produces only 25%.

6 FINAL COMMENTS

Olivier de Schutter, the Special Rapporteur on the right to food from the United Nations, recently said: *It is necessary to consume less energy, to use fewer automobiles and not to create the expectation on the capacity of new technologies to allow us to keep our level of occidental life.* This statement is certainly shocking in some respects, and it is not possible to predict for the future the solution to all the problems related to the supply of food for every citizen around the world.

One concern that should be highlighted is the possible competition between the production of biofuels and food; however, it seems clear that this is not a problem today. In Brazil there has been an expansion in the production of grains in the past few years, and the production of ethanol has also been increasing. It is clear that biofuels should be produced in the regions where the demand is highest, and this is the case in Brazil, where ethanol is largely produced in the south-east of the country. It is also true that with the growth of sugar cane in this area there is a natural displacement of other crops to more distant regions. However, it should not be forgotten that in São Paulo State, for example, crop rotation is practised, meaning that over a period of time other crops like peanuts and soy replace sugar cane.

The development of new technologies for the more efficient production of both food and biofuels is a challenge for all of us, and it should be emphasized that chemistry plays a very special role in the discovery of new alternatives in this field.

References

1 http://www.dryden.ca/website.nsf/museum?OpenPage, accessed in September 2008.
2 www.pfaf.org, accessed in September 2008.
3 C.M. Rezende, V.F.S. Corrêa, A.V.M. Costa, B.C.V. Castro and R.J.V. Alves, *Quim. Nova*, 2004, **27**, 414.
4 www.mme.gov.br, accessed in September 2008.
5 www.tecbio.com.br, accessed in May 2008
6 L.R. Lynd, M.S. Laser, D. Bransby, B.E Dale, B. Davison, R. Hamilton, M. Himmel, M. Keller, J.D McMillan, J. Sheehan and C.E. Wyman, *Nature Biotechnol.*, 2008, **26**, 169.
7 J.D. Keasling and H. Chou, *Nature Biotechnol.*, 2008, **26**, 298.
8 Y.R. Leshkov, C.J. Barrett, Z.Y. Liu and J.A. Dumesic, *Nature*, 2007, **447**, 982.
9 H. Zhao, J.E. Holladay, H. Brown and Z.C. Zhang, *Science*, 2007 **316,** 1597.
10 B. Dale, *J. Agric. Food Chem.* 2008, **56**, 3885.
11 J. Johnson, *Chem. & Eng News*, 2008, **86**, 10.

MUNICIPAL SOLID WASTE AS A RESOURCE RATHER THAN A WASTE. A PRELIMINARY STUDY OF BIOMASS FROM GITHURAI MARKET, AN OPEN MARKET IN THE OUTSKIRTS OF NAIROBI CITY

N. Muriithi,[1]* C. Soi[1] and J.M. Keriko[2]

[1]Chemistry Department, Kenyatta University, P.O. Box 43844-00100 NAIROBI
[2]Chemistry Department, Jomo Kenyatta University P.O. Box 62000-100 NAIROBI

1 INTRODUCTION.

Municipal Solid Waste (MSW) is generated as a result of various human activities in municipalities and cities. The use of the term "waste" is rather inappropriate because several economically useful materials can be recovered from MSW. Examples of such materials include ferrous and non-ferrous metals, polythenes, broken glass, lead batteries, paper, bones, and fruit and vegetable peelings. Materials of organic origin are referred to as biomass.

There are two main problems in recovering economically useful materials from MSW. First, MSW is found as a highly heterogeneous mixture which is extremely hard to sort. At least, this is the case in the city of Nairobi, even in the area in which I live. Second, the materials are not found in sufficient quantities in a particular place. Rather, they are thinly spread over a very wide area and, in most cases, some of the places are not easily accessible because of poor roads or none at all. Consequently, private garbage collectors do not collect garbage from such places. In any case, most of residents in these areas cannot afford the collection fees charged by private garbage collectors.

Biomass is recycled into useful products in three ways. First, easily decomposable materials such as fruit and vegetable peels are made into composite manure. A second method of recycling biomass is to dry it and heat it in the absence of air to make charcoal. Currently, two companies in Nairobi, namely the Kenya Planters and Co-operative Union (KPCU)[1] and Chardust Ltd.,[2] are making charcoal from biomass. KPCU uses coffee husks.

A third method of recycling biomass is to dry it and burn it to generate steam at very high pressure. The steam thus generated is used to turn turbines to generate electricity.[3] In Kenya, the Mumias Sugar Company[4] is generating electricity by burning bagasse. The company actually generates more electricity than it requires for its operations and sells the excess to the Kenya Power and Lighting Company.

When biomass is burnt, the ashes are buried as a waste. Unfortunately, when it rains, heavy metals in the ashes are washed down and thus contaminate ground water.[5] The problem is particularly bad when the biomass is mixed with industrial wastes with metallic components.

**Author to whom enquiries may be addressed.*

2 METHODS AND RESULTS

2.1 Collection and ashing of samples, and objectives

Most of the biomass samples were collected from Githurai market, which is approximately 15km from the centre of Nairobi City along Nairobi-Thika Highway. Coffee husks were obtained from the Kenya Planters Cooperative Union, and rice husks were obtained from rice mills in Mwea Area, Kirinyaga District. Samples with soil particles were washed to remove any soil, the water was wiped off, the samples were left in the open air for any moisture to evaporate, and were then weighed. The samples were then dried at 110°C for at least five hours, cooled in a desiccator and weighed. The samples were then ashed at 550°C, cooled in a desiccator and weighed. The ashes were analyzed for various chemical elements using the X-ray fluorescence spectroscopic technique. The aim of analyzing the ashes was to find out whether any useful components which could be of economic value were present in sufficient quantities.

2.2 Results and discussion

When biomasses containing a lot of moisture, such as banana or pineapple peels, were dried at 110°C, only about 10% of the original mass was left. An important implication of this observation is that in the transportation of such biomasses, a lot of money is actually used to transport water. When the ashes were put in water, some materials dissolved, making the solution strongly alkaline. This was interpreted as indicating the presence of both sodium carbonate and potassium carbonate in the ashes, an observation that was confirmed when quantitative analyses were done. The presence of these elements as carbonates is readily confirmed by treating the ashes with dilute hydrochloric acid, whereupon a strong effervescence of carbon dioxide takes place.

The quantitative analysis data are given in Table 1. A consideration of the data in Table 1 follows. Ashes of the biomasses analyzed contain most of the elements known to be essential for both plant and animal life.[6] Thus:

(i) Our analyses show that the samples contain high levels of four of the six essential plant macronutrients, namely magnesium (MgO), phosphorus (P_2O_5), potassium (K_2O) and calcium (CaO).[7] The other two essential macronutrients are nitrogen and sulphur, which are usually lost as NO_2 and SO_2 when the biomass is ashed, and hence they are not expected among the elements listed in Table 1. Thus, very high levels of potassium and phosphorus in some samples gives a picture of the sample as a fertilizer rather than a waste!

(ii) Our analyses also show that two of the seven essential plant micronutrients are present, namely manganese (MnO) and iron (Fe_2O_3).[7] The other five, namely cobalt, boron, molybdenum, zinc and copper, may be present at very low levels which were below the limits of detection. By "essential", it is understood that, in the absence of this element, a plant will either not grow or, if it grows, it will have a lot of defects (for example, it may not even flower).

In the case of animals, the absence or very low level of an essential element may lead to poor development, sickness and, in case of total absence, to death.[8] For example, if the diet of a chicken is deficient in manganese, its eggs do not hatch,[9] and manganese deficiency is also said to lead to sterility.[9] This is the reason why these elements are deliberately added in feed formulations.[10]

Table 1 *Elemental analysis data obtained by X-Ray Fluorescence Spectroscopy, showing the percentages of the elements when expressed as the oxides shown*

Sample No.	Na_2O	MgO	MnO	P_2O_5	K_2O	CaO	SiO_2	Al_2O_3	TiO_2	Fe_2O_3
1	7.5	1.5	trace	2.6	33.7	28.0	0.1	1.5	0.3	0.5
2	14.7	4.9	0.2	3.8	40.5	12.8	4.5	1.6	0.2	1.7
3	8.3	0.4	0.5	5.1	43.9	2.7	7.5	1.2	0.1	1.0
4	bd	bd	0.1	bd	1.0	1.4	97.4	3.0	Bd	0.4
5	6.5	trace	0.2	2.7	59.1	2.5	3.3	1.8	0.1	0.4
6	14.6	1.8	trace	1.5	38.0	13.3	0.1	2.0	Bd	0.5
7	10.5	1.3	0.1	7.2	56.5	2.6	trace	1.4	0.1	1.3
8	1.9	0.6	2.6	1.6	9.8	8.7	61.6	2.0	trace	0.7
9	6.4	1.9	0.3	0.9	7.0	40.1	5.2	2.3	0.3	2.3

Key: 1, orange peels; 2, pigeon pea pods; 3, pineapple peels; 4, rice husks; 5, ripe banana peels; 6, kale stalks; 7, potato peels; 8, banana leaves; 9, onion stock wastes; bd = below detection.

All the elements shown as their oxides in Table 1, with the exception of silicon, aluminium and titanium, are essential for healthy animal development. In addition, the elements iodine and selenium are required. The current knowledge of processing of minerals is so advanced that the East Rand Gold and Uranium Company recovers gold from materials where its concentration is as little as 0.53ppm.[11] Lithium is recovered economically from spodumene where it is present only to the extent of 1-3%.[12]

The analytical data show that some ashes contain relatively high levels of many essential elements which can, therefore, be recovered economically if desired. For example, the oxides of sodium and potassium are present at very high levels in some of the biomaterials. Thus, the data show that orange peels and banana peels have combined alkalis of about 41% and 65%, respectively.

As is well known, ashes have been used for soap-making for many centuries. Thus, in Babylon, there are records dating back to 2800 BC.[13] No doubt, this was possible because of alkalis in the ashes.

There are two other reasons why alkalis are important. First, in mineral processing, minerals such as chromite[14] require fusing with an alkali to convert the element being extracted into a soluble form. Both sodium carbonate and potassium carbonate are used as fusion mixtures. Second, in the manufacture of glass, silica is fused with either soda ash[15] or potassium carbonate. In our view, ashes with a very high alkali content, such as those of banana wastes, can be used in place of the commercially-available alkalis. Some tests are under way to confirm this hypothesis.

3 CONCLUSIONS

This study has shown that the ashes which are currently treated as a waste are, in fact, an important source of many essential elements for both plant and animal life. For example, and as noted above, studies are underway to find out whether the alkali levels are high enough to replace commercially-available soda ash and potassium carbonate in some mineral processing industries where alkali fusion is necessary. Our analytical data show that some municipal biomasses currently being discarded as waste are an excellent source of alkalis. The current practice of burying the ashes as a waste only increases environmental pollution. Our study suggests that they can be put to better use.

Acknowledgements

One of us (C.S.) is grateful to Kenyatta University for support with research facilities. We are grateful to the Department of Geology and Mines, Ministry of Environment and Natural Resources, for permission to use their research facilities.

References

1 Kenya Planters and Co-operative Union (KPCU) Ltd., Box 72309-00200, Nairobi. Tel: 254 20 210288, E-mail: md@kpcu.co.ke, Dandora Branch (where the coffee coal is made) Tel: 254 20 781910.
2 Chardust Ltd., Box 24371, Nairobi, website www.chardust.com.
3 Energy Information Analysis. *Analyispaper/biomass/pdf/biomass.*
4 Mumias Sugar Company Ltd., listed in Google site under this name.
5 P.T. Williams, Pollutants from Incineration, in "Waste Incineration and the Environment", edited by R.E. Hester and R.M. Harrison, The Royal Society of Chemistry, 1994, pages 27-52.
6 J.J.R. Frausto da Silva and R.J.P. Williams, "The Biological Chemistry of Life", 1997, Oxford University Press Oxford and Academic Press, New York.
7 G.W. Cooke, "Fertilizing for Maximum Yields", English Language Book Society (London), 2nd Ed., 1975, Chapter 9.
8 W. Kaim and B. Schwederski, "Bioinorganic Chemistry: The Inorganic Elements in the Chemistry of Life" John Wiley and Sons, New York, Brisbane, Toronto and Singapore, 1998, Chapter 2.
9 A.C. Deb, "Fundamentals of Biochemistry", 5th Ed., 1992, New Central Book Agency (P) Ltd., Calcutta, pages 384-385.
10 P. McDonald, R.A. Edwards and J.F.D. Greenhalgh, "Animal Nutrition" 4th Ed 1988, Longman Scientific and Technical, Longman House, Burnt Hill, Harlow, Essex, England; and John Wiley and Sons Inc., New York, Appendix Tables 7-14, pages 499-523.
11 B.A. Wills, "Mineral Processing Technology" Butterworth-Heinemann, London, 7th Ed., 2006, page 7.
12 N.N. Greenwood and A. Earnshaw, "Chemistry of the Elements", Elsevier-Amsterdam, Butterworth-Heinemann, London and New York, 2nd Ed., 1997, Chapter 4.
13 http://www.alcasoft.com/soapfact/history.html.
14 S.J. Johnstone and M.G. Johnstone "Minerals for the Chemical Industry", Chapman and Hall, London, 2nd Ed., 1961, page 125.
15 R.N. Shreve and J.A. Brink, Jr., "Chemical Process Industries", McGraw Hill, Tokyo, London, Sydney and New Delhi, 4th Ed., 1977, Chapter 11, pages 179-197.

Subject Index